Successful Service Design for Telecommunications

Successful Service Design for Telecommunications

A comprehensive guide to design and implementation

Sauming Pang
Infosys Technologies Ltd, UK

A John Wiley and Sons, Ltd, Publication

This edition first published 2009
© 2009 John Wiley & Sons Ltd.

Registered office
John Wiley & Sons Ltd, The Atrium, Southern Gate, Chichester, West Sussex, PO19 8SQ,
United Kingdom

For details of our global editorial offices, for customer services and for information about how to apply for
permission to reuse the copyright material in this book please see our website at www.wiley.com.

Library of Congress Cataloging-in-Publication Data

Pang, Sauming.
  Successful service design for telecommunications : a comprehensive guide to design
  and implementation / Sauming Pang
     p. cm.
  Includes bibliographical references and index.
  ISBN 978-0-470-75393-4 (cloth)
  1. Telecommunication—United States. 2. Telecommunication—Management—United States.
  I. Title.
  TK5101.P29 2009
  621.382068′5—dc22

                            2008034405

A catalogue record for this book is available from the British Library.

ISBN 978-0-470-75393-4 (H/B)

Set in 9/11 pt Times by Integra Software Services Pvt. Ltd. Pondicherry, India
Printed in Singapore by Markono

To the memory of my late mother

Contents

Acknowledgements

I would like to thank Professor Gladys Tang, Dr Tony Judge and Richard Newton for encouraging me to write this book. Without them, I never would have started.

I would also like to thank Rod Hart for making this book more readable.

Last, but not the least, I need to thank Philomena Skeffington, who created the Service Design discipline at Energis Communications Ltd.

1

Introduction

1.1 Introduction

The idea of writing this book came from the frustration of trying to describe what I do as a *service designer* and what *service design* is about. Even to the professionals within the telecommunication and service provider sectors (i.e. the companies that provide telecommunication/Internet services), it is often difficult to articulate the concept and purpose of service design. However, I have seen so many projects and service developments fail because there were no service designers in the project team.

Service designers are the people who have the skills to consider all aspects of a service during the design stage. They see how the solution works as a whole (i.e. the end-to-end service), ensuring that nothing is missing from the solution. Service design skills require one to look at the service solution holistically, with the customer experience at the heart of the design. Service design skills are greatly in demand, and yet they are very poorly understood. This book sets out to change that.

The technology sector in general and the telecommunication companies in particular operate in increasingly competitive environments. The companies that survive and excel are those offering the most compelling range of products and services. Since the underlying technologies tend to offer similar features and functionalities, the only differentiation is the services created from these technologies. The method used to create a great service is service design.

There are many books on the market relating to new product development (NPD). Most of them are based on manufacturing products that do not need to be monitored and maintained (or require little monitoring and maintenance), once sold to the customers. To design and introduce a service that needs after-sales care with the maintenance of different technologies is a completely different ball game, and not many people recognize the challenges.

Providing services is where your customers perceive they are served. Your customers normally feel that they are served by human actions, rather than features of your service. In this technology age, feature and functionalities have been taken for granted. It is the human actions or the systems/applications that enable the human actions that make the difference.

The major difference between service design and product development is the concept of providing a service. Designing in capabilities for monitoring and maintaining the service performance and designing the facilities to deliver after-sales support is much more complicated than designing a product. This is especially important for managed telecommunications services and managed information technology (IT) service solutions, where the service providers are expected to resolve the problems when they arise or to fix faults before the customers or end users realize there is a fault.

Successful Service Design for Telecommunications   Sauming Pang
© 2009 John Wiley & Sons, Ltd

A good service is supported by well thought through operational processes, well-designed support systems and good underlying network technologies. The service design skill is the expertise to create and enable such service-supporting functions.

By introducing the concepts of service designer and the service design process described in this book, you can accelerate the process for service launches, increase revenue opportunities, rapidly create new service ideas and reduce design flaws within services. With the reduced risk of design flaws, you will achieve a better service quality, enhance customer service satisfaction and reduce operational cost. The introduction of service designers (in some organizations they are known as technical product managers) will also relieve product managers from internal design work and allow them to concentrate on the market and market opportunities, in turn increasing revenue potential for the services.

Launching new services will introduce changes to the organization. Therefore, when operating the service design process described in this book (as described in Chapter 4), you are also managing the changes to be introduced as a result of the new services. This company-wide process will ensure successful introduction of services into the stable operational environment.

This is the first book on the market that fully defines service design in the telecommunications and service provider environment. Although there is much need for service design skills, it is amazing to see that the concept, the process and potential issues are so undervalued and poorly defined. Working as a service design consultant for many years, I have developed a working model that will provide a consistent approach to designing a scalable and operable service.

This book is technology independent. You can apply the principles laid down in this book to the technology context of the service you are designing. This book can also be used when designing other technological/IT-based services. Readers can adapt the concepts and principles easily to their specific technological environments. In fact, the methodology of service design can be applied to other services that are not related to the telecommunications/IT sectors. The fundamental components of a service (as described in Chapter 6) are much the same, whatever the given context.

1.1.1 Who is this Book for?

This book is written for

- service designers/technical product managers;
- IT professionals;
- network designers;
- product managers/business managers who are involved in defining and managing a service from a business perspective;
- program/project managers;
- system integrators/system integration managers;
- process re-engineering professionals;
- operational managers;
- service managers;
- anyone who is involved in operating services or introducing new services to an operational environment.

1.1.2 Structure of the Book and Who Should Read Which Chapter

Chapter 2 gives the definition of a service and the role of service design. This chapter should be read by everyone (e.g. product managers, program managers, IT professionals, network designers and operational managers) who is involved in the processes of designing and launching new services.

Chapter 3 provides many questions with a view to assisting readers to define business and service requirements. This chapter is designed for business/product managers who are writing service definitions for a service. It also inspires service designers with questions that they should seek answers to in order to capture a more complete set of service requirements at the start of the service design process.

Chapter 4 details the service design process. This should be read by business/product managers and program managers who will be running the service design process to deliver and launch a new service.

Chapter 5 describes, in detail, what is to be done at each stage of the service design process. This chapter is designed for service designers, network designers, IT professionals, system designers, operational personnel, system integrators and program and project managers that are working within the service design process.

Chapter 6 establishes the constituent parts of the service, known as the *service building blocks*. This chapter introduces the reader to all the building blocks required to build a service. It also highlights the factors to be considered when designing a service. This chapter should be read by service designers, IT professionals, network designers and any manager in the telecommunications industry to gain a basic understanding of designing a telecommunication service.

Chapter 7 is a brief introduction to network design. This chapter is written for service designers or professionals who are not familiar with telecommunications network design. It gives readers an appreciation of network design considerations when designing a service. Network designs are normally produced by network designers, but it is useful to have a good comprehension of the component networks in a service, items to be included, design consideration and things to look out for in a network design.

Chapter 8 details the system functional areas for a service and the functions to be considered within each area. By the length of this chapter, you should realize that the systems area is the single most complicated area requiring design. Contrary to popular belief, when designing a service, the systems area requires as much, if not more, consideration than the network area. To achieve 'low touch' efficient operations, support systems play a very important role. The chapter is for service designers as well as IT professionals who are designing systems to support services in an operator and service provider environment. This chapter also give the system integrators/system integration manager an appreciation of the systems to be integrated to form a service.

Chapter 9 describes the necessary operational processes for new service operations. This chapter concentrates on what operational processes need to be defined rather than stating the processes themselves, as these may vary between services and operational organizations. End-to-end operational process examples are, however, given to illustrate the tasks and the sequence of events which need to occur for a cross-section of services. This chapter is written for operational managers and process re-engineering professionals. Since the system functions are closely linked with operational processes, it is also a recommended read for the IT professional designing the support systems.

Chapter 10 introduces the reader to the concept of implementation strategy. It explains how the implementation strategies for a new service can be formulated and which of the implementation strategies can be used for different scenarios. The service solution example given in this chapter provides the program managers, project managers, system integration managers and senior managers with a flavor of the tasks and activities involved when implementing a new service solution.

Chapter 11 explains how the service can be integrated and introduced into an operational environment successfully. This chapter states the stages of service integration and the activities to be performed when launching a service from an operational point of view. It is written for program managers, project managers, operational managers, service managers and system integration and test managers.

Chapter 12 discusses the process and activities to be performed for service withdrawal and service migration. Illustrated examples for both service withdrawal and service migration are given. Key

Table 1.1 Quick reference to using this book[a].

Readers	Chapter											
	1	2	3	4	5	6	7	8	9	10	11	12
Service designers/technical product managers	●	●	●	●	●	●	●	●	●	□	□	●
IT professionals	●	●	□	□	●	●	□	●	●	□	□	●
Network designers	●	●	□	□	●	●	—	□	□	□	□	●
Product managers/business managers	●	●	●	●	□	●	—	—	—	—	—	●
Programme/project managers	●	●	□	●	●	●	□	□	□	●	●	●
System integrators/system integration managers	●	●	—	●	●	●	□	●	□	●	●	●
Process re-engineering professionals	●	●	—	●	●	●	□	●	●	□	□	●
Operational managers	●	●	—	□	●	●	□	●	●	□	●	●
Service managers	●	●	—	□	—	●	—	—	●	□	●	●
Senior managers or manager involved in new service introductions	●	●	□	●	●	●	—	—	—	●	●	●

[a]Chapters for priority reading are marked with '●'; chapters for background reading are marked '□'.

elements for service termination are also highlighted in this chapter. This chapter should be read by service designers, IT professionals, network designers, program managers, business/product managers, operational managers, service managers and any person involved in service withdrawal and service migration activities.

Table 1.1 gives a quick reference guide for the different professionals using this book.

1.1.3 Definitions

This book uses a variety of specialist terminology. The most common specialist terms are defined in the following list to help the reader gain the most benefit from the rest of the book.

Cancellation. Discontinue the service contract before service provisioning is complete.

Customers. Customers are organizations or businesses who buy services. For a telecommunication operator who serves business/enterprise customers and wholesalers, the customers are mainly the corporations who buy end-to-end solutions from the operators or they are service providers who buy wholesale solutions from the operators. For an operator who services retail customers, it is the retail customer who buys the service directly from the operator. The retail customers, therefore, are defined as end users for the purpose of this book.

End users. End users are users of the service. In a telecommunication operator environment that serves business customers, these are mainly the employees who use the service provided by the operators; for example, a sales team member (the end user) may use the third-generation (3G) remote access service provided by an operator to access their e-mail and IT applications. In a service provider environment, these are users that have service contracts with the service providers, but part of the solution/service may be provided by an operator. For example, for subscribers to an Internet service providers (ISP's) broadband service, the subscribers are the end users. They may use part of the wholesale broadband capabilities provided to the ISP by a telecommunication operator, but the ISPs may provide additional value-added services on top of the broadband capabilities.

Network element. A piece of network equipment that supports certain network functions required by the service.

Network node. A collection of network elements that are physically located together at one site. They do not necessary support the same service.

Network termination. This occurs when the customers/end users decide to cease or terminate the network connectivity. This can be a result of discontinuing the service contract after service provisioning is complete (i.e. terminating the service) or the network connection is no longer required (e.g. moving out of the current location). Most of the time, these terminations are initiated by the customer/end user.

Service migration. This involves migrating customers/end users from one service to another. This could take place within an operator/service provider or between operator/service providers.

Service termination/service ceasing. This occurs when the customers/end users discontinue the service contract after service provisioning is complete. Most of the time, these terminations are initiated by the customer/end users. Sometimes this is initiated by the service provider/operator when the customers/end users fail to pay their bills.

Service withdrawal. This is withdrawing the service from use. This occurs when the service comes to the natural end of its useful life, mainly due to commercial or technological changes.

System users. System users are the users of the business support systems or operational support systems (OSSs). Typically, they are the operational personnel supporting the service.

Network node. A collection of network elements that are operated by the same owner, which do not necessarily support the same service.

Network termination. This occurs when the customer reaches the point of termination of network connectivity.

2

What is Service Design?

This chapter puts *service design* in context. It establishes the definition of a service and the role of *service designers*. It also highlights the differences between services, products and applications, as well as intra-domain and inter-domain services.

2.1 What is a Service?

We use the word 'service' all the time, but it is hard to find a good definition. In the *Oxford Advanced Learner's Dictionary* [1], the definitions of service are 'a system that provides something that the public needs, organized by the government or a private company' or 'a business whose work involves doing something for customers but not producing goods; the work that such business does: financial/banking/insurance services'.

My definition of a service is: something a business or person provides to customers of which the most important elements are often intangible and the quality of the service is subjective and cannot be measured easily.

In an increasingly competitive telecommunication environment, it is the service that really matters. Most organizations understand that winning customers costs much more than retaining them. With good services, it is likely that the customers will continue to buy your services and recommend your services to their friends. You certainly do not want your customers/end users to walk away from the service because they feel it is a bad service. Perception of a good service and good value for money are often what matters, rather than, for instance, the wonderful network features on offer.

Most service providers will offer enticements to prospective customers by listing a set of features that they might want (e.g. 300 min talk time per month; download capability of 2 Mbits/s; free voice mail). Their intention is, of course, to make the customers feel that they need something tangible that they can/will use.

Trying to make your customers feel good about your service is probably one of the most important, but difficult, things to do. How do you convince your customers that you are providing a good service and that you are better than your competitor? How do you demonstrate to your customers that you are providing a good service? Quantifying and measuring the quality of a service is a complicated process.

Most service providers will provide business customers with a set of service levels in an agreement (referred to universally as a service-level agreement (SLA)) when signing the contract (e.g. a fault

Successful Service Design for Telecommunications   Sauming Pang
© 2009 John Wiley & Sons, Ltd

will be fixed within 3 days of reporting it; new service will be provisioned within 2 days). These SLAs should measure how good or bad that the service is. An SLA is a vehicle where the service provider can demonstrate to their customers that they are providing a good service. These are the tangible measures that help the service providers to quantify how well they have been looking after their customers and how well looked after customers might feel when using the service. In theory, if all the SLAs (provided that they are meaningful) are met, the customers should feel they are getting good service. However, in practice, it is much more personal than that. It is the relationships the service providers have with their customers that really matter.

The front-line customer facing representatives need to project the image of your company and the way in which they handle customer enquiries and complaints. These are the experiences that the customers will remember most about the service. Most customers nowadays take it for granted that the technology should work well, and they will only remember the technology if it is very clever or impressive or if it does not work. It is the 'personal' response to their requests (e.g. orders, fault reports, complaints or change requests) that creates the competitive differentiation between services. By 'personal' response, I do not mean that you need to have people responding to all the requests all the time – that will be too costly, but the responses should have as many personal touches as possible. These personal responses are the results of well designed support systems and good operational processes within the service providers.

To provide customers with the service features they want and the mechanism to measure the quality of their services with 'personal' responses, we need to design the service with the customers in mind. The task of designing a good service comprises:

- capturing a complete set of customer, business and service requirements;
- designing and using suitable telecommunication network technologies;
- designing and using suitable IT systems that manage those technologies;
- designing the Business Support Systems (BSSs), the Operational Support Systems (OSSs) and operational processes that perform the customer and service management functions across all parts of the company.

What makes telecommunication services (and, to some extent, large-scale IT services) unique, and different from that of manufactured products, is the complexity involved when designing the service that touches all parts of the company and the need to take into consideration all of the elements listed above.

In this book, I will provide a comprehensive guide to this complex subject. All the different factors to be considered, namely the management and operational processes required for successful service introductions into the operator/service provider environment, will be described in detail.

Note: strategies for maintaining the customer relationships are outside the scope of this book. However, services designed using the methodology in this book will have the tools and information that will facilitate the management of customer relationships.

2.2 What are the Differences Between a Service and a Product?

To understand the differences between a service and a product, it is useful have a definition for a product. In the *Oxford Advanced Learner's Dictionary* [1], the definition of product is 'a thing that is grown or produced, usually for sale: dairy/meat/pharmaceutical products'.

My definition of a product is something (normally physical) a business or person makes (typically from manufacturing) and is capable of being sold many times over with minimal after-sales service (e.g. television, cars). By comparing the definition of product defined here and the definition of a service defined in the previous section, you can start to understand the differences between the two.

Unlike most consumer products (e.g. television), good telecommunications service providers will provide their customers/end users with a service rather than just selling them a product. Selling a telecommunications product is like selling a phone to the end user with no connection to any network and none of the phone features enabled. You will probably remain unsatisfied and are likely to buy the telephone service so that you can use the telephone you have just bought. Once you have bought the telephone service, you will expect the service provider to fix it when broken.

Therefore, the major differences between a product and a service are that a service incorporates the after-sales care, which includes monitoring, maintaining and fixing the different technologies that make up the service, as well as providing facilities to deal with customer enquiries, customer complaints and service requests. In the case of a product (e.g. television), once it is sold, as a seller, you do not need to take an interest when it is broken. However, for a service (e.g. broadband service), you, the seller/service provider, need to monitor and maintain the service. You, the service provider, need to fix it when it is broken. The service providers are also expected to answer any queries regarding the service and to provide technical support for the service, as well as perform any service/change requests as required by the customers. It is the after-sales care that differentiates your service from that of your competitors. Providing a good after-sales service is the mechanism you can use to distinguish yourselves. However, it is the provision of the after-sales service that introduces complexity into the design of a service, and it is also the one that will cost you most!

2.3 Service Versus Network Capabilities

There has been much misconception between technical capabilities and good services. Having the most advanced network capabilities does not always automatically equate to a good service. You may have a network capability to provide your customers with an Internet access speed of 20 Mbits/s utilizing the existing network infrastructure with a minimal incremental cost. Fantastic! However, does that equate to a good service? Well, that depends.

It is all very well to have great network capabilities, but will the customers want it? What are the applications that can use these wonderful network capabilities? What are the services that can use this network capability? Do you have the right pricing strategy in light of this network capability? How about the management of the network elements? How quickly can you provide this service to your customers? How many customer visits do you need to provide this service? Who goes to fix it when it goes wrong? How soon can you resolve a customer-reported fault? How about customer complaints? The answers to these questions will be crucial for designing a good service.

Designing an end-to-end service is about turning the network technologies into something that the customers want to use; making the customers feel that they are being looked after whilst using the network capabilities; providing the capabilities and tools for the operational personnel to looking after the customers and ensuring that customer payments are collected to make the business viable – all at the same time.

Network capabilities certainly drive many new services we have today, however having a 'service wrap' around the network capabilities is just as important. Without the service capabilities, you cannot manage the network capabilities, you cannot fulfill any orders (or the customers cannot order anything) and you cannot handle any faults or complaints. In fact, you will not have any customers at all, never mind keeping them happy.

In the chapters that follow, I will explain how you can turn the network capabilities into manageable and sustainable services.

2.4 What are the Differences Between a Service and an Application?

Services and applications seem to be interchangeable, especially for mobile services. However, there are distinct differences between the two. As mentioned earlier, services are something a business provides that customers want and which make them feel that they are looked after when using the service. Applications are service enablers that customers can use to make certain transactions or to access different services. For example, an Internet banking service will require an application provided by the bank to enable a secure banking transaction. Therefore, the application enables the service, while the service is the transaction itself (in this instance).

The application quite often forms part of the service. In the example of online shopping, the application enables the customers to browse the online shop, select the items to buy and purchase the goods. The service is providing the ability to browse the catalogue of the online shop, make the catalogue of the shop available, check whether the goods are in stock, enable the customers to purchase the desired items, collect payment and deliver the items to the customers on the agreed date.

For most mobile and fixed-line services, it is the applications that drive the usages of the services, rather than the service itself. Consider it as a layered model with the application layer sitting above that of a single service or across multiple services. For example, for a broadband Internet access service, the application for that service can be e-mail, download of content, chatting on social websites and so on. Most applications are software driven and may use different services to achieve the required transactions. These applications may or not may not be provided by the operator/service providers that supply the telecommunication service. Often, these applications are provided by a third party or by content providers, which is causing rapid changes to the customer value chain.

2.5 Intra-Domain Versus Inter-Domain Services

Intra-domain services are normally referred to as using the same technologies, functions or technology platforms. Inter-domain services refer to services that involve multiple services, technology platforms or multiple operators. For example, in 3G mobile services, there is the circuit switched (CS) domain and the packet switched (PS) domain. Services can be within one domain, known as an intra-domain service (e.g. public switch telephone network (PSTN), managed data service), or across different domains, known as an inter-domain service (e.g. Voice over IP (VoIP) service, 3G mobile services). For inter-domain services, there are more technical and operational considerations when designing these services.

2.6 What is Service Design and What is the Role of a Service Designer?

2.6.1 What is Service Design?

Service design is the skill and the methodology used to create great services. It takes a holistic view of the service, taking into account the customers' experiences as well as the technical and operational aspects of the service. It ensures that the end-to-end service solution is completed and that no stone is left unturned. It confirms that everything within the service works harmoniously together as a whole.

As mentioned in Chapter 1, and in the previous sections, it is the service that really differentiates you from the competition. The network technology alone is no longer enough. Service design is the skill that turns network capabilities into manageable and sustainable services. The skill is used when creating new services and enhancing existing services. Therefore, the service design skill used to create distinguished services is much in demand, and it is paramount that you have a service designer in the development team when designing and launching new services.

Many designers, product managers and program managers think they can design a service; however, many fail to consider all aspects of a service during the design stage. As a result, many services are launched haphazardly. Consequently, there are not many great services around. These services typically suffer from high operational costs. There is a marked difference in customer experience and operational efficiency between a great service and one that is cobbled together.

Service design involves:

- capturing a complete set of customer, business and service requirements;
- using suitable telecommunication network technologies to design the service solution;
- designing and using suitable IT systems to manage those technologies;
- designing and using the appropriate OSSs and processes to perform the customer and service management functions across all operational areas;
- ensuring the consistency within the service solution;
- verifying all the constituent parts of the service are in sync.

What makes designing telecommunication services challenging is the complexity involved when designing the service (as it touches all parts of the business) and the amount of changes involved during the birth of a service. The skills and methodology used to engage these challenges are service design. This book unlocks the mystery of service design and the role of service designers.

2.6.2 What is the Role of a Service Designer?

Designing a service is like designing a house. Great architects build good-looking, yet functional, houses. This is equally true for a great service. A lot of thought goes into considering the customer experience, operational efficiency and the 'service wrap' that is around it; just like living in a well-designed house.

What makes designing telecommunication services more difficult than other types of service is the complexity involved when designing the service, as it comprises the telecommunication network technologies, IT systems that manage those technologies and the operational support systems and processes that perform the customer and service management functions. Just designing the network technology is only a small part of the jigsaw. Designing a service touches all parts of the telecommunication business; hence the complexity. Sadly, this is not widely appreciated.

Service designers are the architects of services – the ones with the vision of end user and customer experiences, the ones who decide how the services work end to end and the ones that have the capability of envisaging how the various parts of the jigsaw within the services fit together. Service designers are there to ensure every aspect of the service works harmoniously together and that there is no gap in the service solution. Service designers are also the design authorities for their services.

2.6.3 Product Managers Versus Service Designers

In most telecommunications companies, there are product managers who manage and own the products/services. The product manager's role is to look at the market, spotting new revenue opportunities for the product areas and making the business case viable for new revenue opportunities. These revenue opportunities may come from new customers or a new segment of the market using the same services or may be realized through launching new services. The product managers are marketers who look outward from the company. They should not be the person designing a coherent service solution. That is the role for the service designer. Product marketing is primarily a marketing function, whereas service design is a design/engineering function. Product managers are marketers, not engineers or designers. The mind-set and the skills for both are very different; and yet not many telecommunications companies recognize that. They continue to expect the product managers to perform both roles. Consequently, many services are cobbled together and many operational problems arise as a result. Using the house analogy again, I sometimes do wonder whether anyone would consider asking estate agents to design and architect their house.

3

Service: A Business Perspective

After establishing what a service is and how the customers should feel when using a service, it is useful to go back to the basics and examine what telecommunication services are for. Why do your customers want or need the service? Telecommunication services enable people to communicate with each other and provide a means to access information or content that people need or want. It is also a medium through which different service industries (e.g. retail, travel or financial service) can communicate with each other and their customers. Hence, telecommunication services can be an additional channel to market for their services or applications. Different customers and sectors of other industries have different requirements on telecommunication services; however, all businesses, no matter large or small, need telecommunication services to operate. It is important, therefore, to understand the market you are serving and the customers' needs and requirements before designing the services. Without a good understanding of the customers' requirements, you will be wasting time and effort (and money) designing for services that nobody wants and in turn lose out on potential business opportunities you could have.

Many telecommunication companies are not clear about their business requirements, due to the lack of understanding of the market. This results in the lack of clear requirements or incomplete definition of the service, which then leads to inappropriate or incomplete solutions with designers trying to second guess what is required. That is certainly not a recipe for success. In this chapter, I attempt to inspire you, through many searching questions, to capture a complete set of requirements for the service.

Before going into the technicalities of service design, I would also like to put services in a business perspective. Is there a market for the service? Why does your company want or need new services? Is it because the customers ask for it? Or is it that the new technology can offer lots of features? Or will the new service help to unlock potential markets? Or will the service help to lock in existing and potential customers/end users? Why do the customers/end users want the services? Of course (I hope anyway), the answer is a combination of all the reasons listed above. However, most importantly, launching new services is about exploring new business opportunities, growing the business and revenue and delivering shareholder values (i.e. making money).

So how do we do that without costing the Earth and making the birth of a new service relatively painless, efficient and with minimal risks to the company? The answers to these questions lie within the rest of this book. However, in this chapter, we start by exploring the different activities, preconditions and considerations before launching into a full-blown service design activity. We will also be looking at different constraints and success criteria of a service.

Successful Service Design for Telecommunications   Sauming Pang
© 2009 John Wiley & Sons, Ltd

3.1 Preconditions for Service Design

3.1.1 Market Analysis

Before you even start considering a service concept, one needs to consider whether the new service fits with your company's strategy or business objectives. There is little point in considering a residential/consumer service if your company strategy is purely for corporate/business customers with no wholesale business that will provide services to residential customers/consumers. Once you have established that the service fits the company strategy, you need to decide whether there is a market for the potential service. Therefore, one of the first activities for considering a new service is to perform a marketing analysis, before any design activities should take place.

Market analysis should include (but not limited to):

- Identifying the aim of the service. Is the aim of the new service gaining new market share? Or is it responding to competitors and, hence, to retain current customers? Or is it to become the market leader?
- Identifying market drivers, trends and the customer's need for the service.
- Describing the potential business benefits for potential or current customers by introducing the service (e.g. if the customer uses this service, the potential cost savings for the customer's organization will be x).
- Assessing your company's current service portfolio. How will the new service enhance (or otherwise) the current portfolio? Is this a niche service? If so, what is the leverage to widen the market potential?
- Performing analysis on the strengths, weaknesses, opportunities and threats (SWOT) that your company is facing with (and/or without) the new service. What are the risks and implications of launching or not launching the new service within a certain timescale?
- Identifying potential customer profiles and target market segment(s).
- Calculating the size of potential market (e.g. the potential customer base, the number of potential end users) and geographic availability of the service.
- Estimating the value of the market (e.g. the value of each customer/end user) and margin for each customer/end user.
- Identifying potential additional service requirements from initial service concept.
- Analysing how this service may complement or be bundled or compete with current services being offered within the company. Is this service going to substitute current service(s)? Is this service going to be a rival service to another service that is currently in the portfolio?
- Performing competitive analysis (including the competitor's service offerings, pricing and pricing structure, potential opportunities and risks to your company).
- Estimating the cost of lost opportunities (i.e. the potential opportunity costs) without the service and within what timescale.

This list is not meant to be exhaustive, but these are the main areas to consider before the start of any service design activity/starting the device design process.

3.1.2 Business Case Development and Pricing

The next to step is to build the business case for the service, once the business and marketing proposition of the service have been established. Even if the service has massive market potential, if the business case does not make sense (i.e. the service does not make money with reasonable pricing or the return on investment (ROI) is over a very long time), then starting the service design process becomes questionable, unless for business strategic reasons.

To develop a business case, you will need cost input. The development cost, mainly comprised of one-off costs (e.g. cost of development team, cost of buying, building, installing and testing network and systems, cost of training) and recurring expenditure (e.g. cost of leasing buildings to host the equipment, cost of customer premise equipment (CPE)/network termination devices). Other cost items may include the cost of sales, cost of marketing the service and operational cost for supporting the service (e.g. cost of call centers dealing with enquiries, billing complaints, fault reporting and resolution). These costs are normally categorized as capital expenditure (CAPEX) and operational costs (OPEX). The categorizing of different cost items depends on the accounting practice of your company. The cost of developing the service and the ongoing support cost of the service are obtained through a feasibility study of the service (please see Section 5.2 for details). Other costs for consideration may also include the cost of withdrawing the service and opportunity gain at the expense of another service that the new one is substituting.

After examining the competitor's pricing (if this exists), the cost of the service (both CAPEX and OPEX) and the pricing of other services within the company's portfolio, you should be in the position to determine the price, pricing structure and potential revenue for the service that will make the business case work in your favor.

The pricing structure of the service has major implications on the billing systems. Therefore, when designing the pricing structure, it may be a good idea to investigate the current billing capabilities of your company. This can, of course, be part of the feasibility study to be conducted. However, if the new pricing structure requires a new billing system or new module within the billing system, then it may be a costly affair and may not be the most cost-effective solution, never mind the lead time for implementation, which may say goodbye to the service opportunity. So, my advice is to keep things simple.

3.1.3 Service Description

The service description (sometimes known as the service definition) is the most important document as far as the service requirements are concerned. It defines

- what the service is;
- the drivers for having and developing the service;
- what service features are on offer;
- customer experience when using the service;
- how the customers/end users will use the service feature and the benefits for each service feature;
- what service levels the service need to support, and so on.

The service description also identifies the business and marketing requirements for the service. The service description should state all the service features/features combinations that are allowed within the service framework and the pricing associated with these combinations. It is important to capture these service variants within the service description, because certain feature combinations may not be technically possible or very expensive to implement. If these are not documented, then the sales team may sell things your company cannot do or are very expensive to implement. This will avoid unnecessary confusion or customer requests after service launch. The service description should remain an internal document and be separated from the marketing collateral that is created for external (customers/end users) use.

In the sections that follow, I will take you through all the different areas of a service, from a business and service perspective, with the view to helping identify all the requirements that a service description should contain. The service description template in Section 3.12 will provide a summary of items to be included.

3.1.4 Requirements Definition

Before launching into a discussion about a service and business requirements definition, it is worth spending a little time on understanding what requirements are and why we need them. Requirements define what is needed or what the customers/end users want. They specify certain criteria to be met and sometimes the context in which these requirements are used or to be met.

Why do we need requirements? We need requirements because it is virtually impossible to design anything without any requirements. If you have not specified what is required, then you will get what is given and it will probably not be fit for purpose. If you really think about it, how can you expect someone to design or delivery anything if they do not know what it is that they are supposed to design or deliver?

Many people also think that writing an exhaustive list of requirements is a waste of time. By the time the requirements are written, things have moved on and what is written is no longer relevant. Why bother writing it? Yes, writing an exhaustive list of requirements is almost impossible, but without any requirements or with requirements that are not complete it is also impossible to design a solution that is complete. Having defined requirements is also a very good way to keep track of and control what is being delivered. Managing changing requirements seems to be a moving target, but is essential to ensure the project is under control.

3.1.4.1 Capturing Requirements

When was the last time that you asked someone for requirements and the answer was: 'What have you got'? or 'I want this solution.' or 'What's your problem, why don't you just do it'? These answers are hardly useful from a design point of view and are some common mistakes most people make at the start of the service design process or at the start of any project in general. Designing a service or delivering a project is not like going into a car showroom. You cannot just pick something that you like the look of off the shelf. I suppose you can, but you still need at least a list of criteria by which you measure your choice against. These are your requirements. Capturing requirements is a black art for many, but absolutely essential to the success of designing a service.

From past experience, not many people know what they want; hence, they do not know their requirements. One of the well-known techniques for capturing requirements is asking open questions that start with 'what', 'how', 'why' and 'who' and so on. Closed questions are only good for confirmation. When capturing requirements, open questions may lead you to more answers. Further questions will arise from some of the answers, either to clarify the requirements or additional requirements. Making assumptions based on the answers to your questions as to what people might want is also a bad idea. It is much better to clarify things you are not sure about. Often, this may open a completely new area that you have never considered and that requires further exploration.

Since requirements are normally quite abstract, another way of confirming/capturing detail requirements may involve building a prototype. Having a 'rough and ready' model gives the person specifying the requirements something tangible to look at. This will help the person specifying the requirements to think about what they really want and, in turn, further refine their requirements. Having a prototype also clarifies what they want the service/system to do and how (steps to go through) to perform certain tasks. This is especially useful for specifying requirements for systems and applications.

3.1.4.2 Attributes of Good Requirements

Everyone tends to assume that writing requirements is easy, but it is not. Writing unambiguous, measurable and testable requirements that truly represent what you need is not as straightforward as you think. Writing testable requirements can involve the definitions of many parameters and require a full understanding of context and variables. Most people hugely underestimate the importance of requirements and the effort required to define them.

Having requirements that are not testable or measurable is not very useful. You may as well not have them at all. How do you know if the solution that has been designed and delivered is what you wanted (i.e. fulfills your requirements)? The only way is to measure and verify against it (i.e. test it). If the requirements are not testable, how do you verify if the solution delivers the service that is required by the market? For example, you cannot have a requirement saying that the solution shall be scalable. This is not testable. You must define how scalable (e.g. support up to 10 000 simultaneous concurrent sessions with throughput of x).

When defining requirements, the prescribed solution quite often becomes the requirement. For example, in a broadband service, you can have a requirement of 'the service shall be delivered over asynchronous digital subscriber loop (ADSL) technology'. Well, ADSL technology is really a solution and not a requirement. The requirement should state 'the access network cost per end user of the service shall be on or below $\$xx$ and has a geographic coverage of $x\%$ of the population in the country with an access speed of x Mbit/s'. As you can see, stating the requirements objectively and analyzing them is a nontrivial task. The requirements should state what are required, rather than how they are achieved or delivered.

Most people define requirements as functions and features, but there is a whole lot more to requirements than that. For a start, think about nonfunctional requirements. What is the response time of the system if you ask it to perform a certain function under a certain load? There is little use for the system users if it takes 10 min to perform a certain function. I guess this will give the system users a good excuse to make a cup of tea! However, this is not a very efficient use of time, and if all the functions in the system take this long then there can be only so many cups of tea a system user can drink before getting really fed up! The other area for consideration for nonfunctional requirements is the parameters and associated boundaries. For example, there could be a requirement stating 'the network availability shall be 99.9%', but over what timescale? Is it measured on a per month basis or is it an average over a year? If it is on a daily basis, then it is probably a requirement that is impossible to achieve. Another area of nonfunctional requirements to think about is usability requirements. The feature will become absolutely unusable if the end user needs to press 10 different buttons on the mobile phone to use it. Therefore, it is important to consider nonfunctional requirements rather than just features and functionalities.

Consistency is another attribute for defining good requirements. In the requirements context, consistency is to ensure that none of the requirements conflict with each other. This normally happens if one group of users wants one thing and other groups want something slightly different. For example, the first-line support personnel from the network management center want a particular network alarm to be marked as 'critical' in the network management system, while the second-line support personnel only want to see that alarm as 'major'. Getting different user groups to agree on requirements is definitely not an easy task. However, conflicting requirements are not acceptable either.

Last of all, requirements should be realistic. There is little point in having requirements that are never going to be fulfilled. If there is a vision where the service should get to, then this should not be captured as requirements, but rather as information statements.

3.1.4.3 Other Requirements Attributes

When defining and writing requirements, it is useful, sometimes essential, to number the requirements and state the source of the requirements (i.e. where the requirements come from). Numbering the requirements enables the ease of reference for requirements management, compliance statements for designs and test specification coverage. Requirements traceability is the term used for logging the sources of requirements and tracking the changes of the requirements during the development life cycle. Requirements traceability is especially important if you have a large number of requirements and if conflicting requirements exist. It is not good enough to have the source of the requirements to reside in the head of the person writing the requirements, especially if the sources of the requirements result from interviewing many people around the business. It is almost impossible to remember where all the requirements come from. In addition, without knowing the requirement traceability, it will be very difficult for anyone to manage the requirements during the development, because no one will know which part of the requirements have been changed and why the requirements have been changed.

Last, but not the least, is requirements prioritization. For each service, there are many, many requirements. Most of the time, it is almost impossible to fulfill them all. So, which of them are essential (i.e. 'must have')? Without them, there is no service. Some requirements reflect features that are needed but the service will still function without them (i.e. the 'should haves'). Of course, there are some requirements that are 'nice to have'. One has to be ruthless, but consistent, when prioritizing requirements. Most requirements in the first instance are essential; but, if you really question them (e.g. what if this requirement is not fulfilled, what will happen?), quite often they are not as essential as one initially might think.

As you can see, requirements capture, analysis and documentation is a task of its own and is often onerous and dull – but it is so important. One needs to think about all the areas concerned. Ensuring everything is captured and without gaps is not that easy. Most people just assume that writing requirements is something that can be done very easily, whereas it often lies at the root of many problems later on. Writing requirements is an iterative process. No doubt, requirements will change over time. Therefore, it is important to have a good change control and traceability process in place to ensure that the origins of the requirements are clear, that changes are tracked and that the requirements are managed effectively. It is recommended that requirements capturing tools are used to log all the requirements and their priorities. This will also make requirements traceability, management and change control much easier.

Requirement definition, management and traceability is a subject in its own right. Further reading on this subject can be found in *Software Requirements Styles and Techniques* [20] and *Software Requirements* [17]. The standards for requirements definition can also be found in the *IEEE Recommended Practice for Software Requirements Specifications* [18].

With all of the above in mind, how do you define requirements? Where do you get them from? First of all, for most services, requirements come from the service description, with service operational requirements from around the business. Therefore, having a fully defined service description is so important to the success of developing the service. Secondly, I find that it is very rare that requirements are handed to you on a silver platter; you need to go out and hunt for them. The most effective way of capturing and defining the requirements is to ask the right people lots of open and searching questions. Hence, for the rest of this chapter, I will be asking many questions in order to inspire you to identify and pinpoint potential requirements.

3.2 Business Requirements

Business requirements are one of the key areas for defining a service. Incomplete or unclear business requirements often lead to the lack of or incomplete definition of the service. So what are classed as business requirements as opposed to other requirements? Why do we need to know about business requirements?

Business requirements describe the needs from a business perspective. For example, what are the channels of sales for the service? Is it through the reseller or through direct sales channels, or both? Is this service being provided as part of a bundled service or is it independent from current ones, or both? What are the pricing schemes or commission schemes for the service? What are the SLAs or response times the service needs to support? All of these questions lead to business requirements.

Why do we need to know about business requirements? Business requirements define what the service is, what the service is for and what types of market the service is serving. For example, service levels for business customers will be need to be higher than that of consumer services. This, in turn, forms the basis of some service requirements relating to service features and service quality.

Business requirements also have wide implications for the solution to be designed. Following through with the example above, in order to support a certain SLA (e.g. resolve a fault within 3 h), operations may need to employ a few more people and the agreements with third-party maintenance providers will need to be negotiated with this service level in mind. In addition, there will certainly be reporting requirements associated with this, as the customers will want quantified measures to ensure their required levels of service are met. As you can see, without clear business requirements, the solution for the service will not be complete.

In order to help you to identify and clarify business requirements, there are some questions below (broken down into different areas) that you should consider asking. This is not meant to be an exhaustive list, but should give you a good start with capturing business requirements.

3.2.1 Sales and Marketing Questions

Below are some sales and marketing questions one should be asking.

- What is the target market of this service? What are the market segments this service is addressing?
- What are the benefits of this service to the target market/market segment? What are the benefits to the customers/end users?
- What is the value of the market segment? How many potential customers/end users will be using the service? What is the average revenue per user?
- What is the target margin of this service?
- What is the geographic coverage of this service?
- Is there a launch timetable?
- What are the channels of sales for the service? Is it through the reseller or through direct sales channels, or both?
- What are the pricing schemes or commission schemes for the service?
- What sales personnel are required: just an account manager, or are some technical experts required to support the sales activities?
- What are the sales forecast (in terms of the number of customers and end users) for the first 3 years of the service?
- What are the minimum terms of the contract? What are the charges if a customer terminates the contract before reaching the end of the minimum term of contract? How are these charges calculated? Is there a requirement to generate this charge from a system?
- Is there a minimum commitment on volume per customer/reseller within a defined period? This is particularly important for wholesale service and services that are sold through resellers.
- What are the penalty charges if minimum commitment is not achieved within the defined period?
- Are there penalty charges if the minimum committed volume is not met? This is important for wholesale business.

- Do the customers need to satisfy certain criteria before they are qualified for the service? If so, what are the criteria? This is important for wholesale service and services that are sold through resellers.
- Within the customer engagement process, what technical resources are required?

3.2.2 Pricing, Charging, Billing and Accounting Questions

- What are the chargeable items for the service?
- What is the pricing mechanism for each chargeable item? Is it subscription based or usage based?
- For the usage-based services, how will the usage be measured? Will it be on a time basis or the amount of data transfer (e.g. $x per Mbit of download) or pay per use (e.g. buy a piece of content on one-off basis)?
- For a time-based charging service, will there be different rates for different times of the day and day of week usage (e.g. charge for daytime use is $x per minute, while evenings are $y and weekend calls are free)? This could equally apply to data usage rate, where different amount per Mbit can be charged at different times of the day.
- Will different levels of service be charged differently (e.g. charging different quality of service (QoS) levels at different prices)?
- Will the customers be charged for installation for the service?
- What are the recurring charges for the service? What are the charging periods (e.g. monthly, quarterly or yearly)? Are the recurring charges in advance or in arrears?
- Are there any discount schemes for the service? If so, what are they? Which customers are eligible? Are they based on the amount billed or volume of connection and so on?
- Is there a revenue share scheme for this service? If so, at what point is that revenue share scheme triggered? What criteria do the customers need to fulfill to join the scheme? Is this trigger automatic or does someone need to authorize this? If it is automatically triggered, one needs to consider the business risks for doing so.
- Do the customers have a choice on which day of the month or which month of the year the bills are to be sent out?
- What is the payment term (i.e. number of days the customer needs to pay the bill after billing date) for this service or is it determined on a per customer basis? Or is this a pre-paid service, or can the service be both pre-paid and post-paid?
- What is the billing accuracy required (in terms of guarantee of delivery on time, the accuracy of each bill)?
- What is the customer handling procedure for bill disputes? Who in the business has the authority to authorize any billing charges being written off?
- What is the billing format (i.e. what does the bill need to look like)?
- Is there a requirement for sending out bill payment reminders? If so, under what circumstances and in what form? Do you remind the customer/end user after the bill is overdue for a month? Do you send a 'red bill' or do you call the customer/end user or do you send them a text message? Does anyone in the organization need to be notified? Does the overdue amount matter? If so, what is the amount and what escalation is required?
- What monitoring facilities are required for overdue bill payments?
- What medium will the bills be in (i.e. is the bill electronic, paper)? How are the bills distributed to customers/end users (i.e. bills access via a Web portal, paper bill to be posted or bills with call records on a CD)?
- What are the methods for bill payment?
- How are these bills payment accounted for?

- At what point will the service be disconnected if customers/end users have not been paying their bills?
- For the usage-based services, do call records or usages records need to be provided to the customers/end users? Data concerned could be huge, what medium should the data be presented in? Is there a requirement for a summary to be provided? If so, what is the summary based on?
- Is there a requirement to verify all other license operator (OLO) bills as a result of buying network capacity from another operator or from calls delivered between network operators for this service? What billing items require reconciliation?
- What are the revenue accounting requirements for the service?
- What are the revenue assurance requirements for the service?

3.2.3 Customer Service Support and SLA Questions

- What is the service provisioning lead time (e.g. from contract signature to service delivery)?
- What is the response time for a fault reported by customer?
- What are the definitions of different faults and what are the fix times for each fault categories? For example, category 1 faults may be defined as a fault causing a total loss of service. The guarantee time to fix is within 4 h of a customer reporting the fault.
- Is there a requirement for a customer help desk to deal with customer enquiries (e.g. billing queries or service information enquiries)?
- What kind of customer/end-user support will be provided? Do you need a technical help desk? Will the customers/end users be calling a help desk with installation queries or will they be calling for general product enquiries?
- What are the expected hours of operation for customer/end-user support and services? For example, is it an 09:00–17:30 Mon–Fri operation or is it a 24 h, 7 days a week operation?
- What is the predicted volume of customer services calls?
- How will the customer/end users interact with the service provider/operator regarding this service? What are the scenarios where customers need to contact the service providers/operators apart from general enquiries, ordering the service, reporting a fault, changing or adding service packages, moving location, receiving bills, paying bills and receiving customer reports?
- What is the interface for the customers/end users regarding fault reporting? Is there a fault help line they can ring? Can they report a fault through a portal?
- What is the interface for the customers/end users ordering a service? Through the account manager only or can it be through a portal Web site? If it is through a Web portal, what are the authentication and authorization for these orders?
- What is the interface to the customers/end users regarding move, add and changes? Do they need to write in/call or dynamically make certain changes through a web portal? What are the authentication and authorization for these request?
- Will the service require a remote diagnostic function to diagnose customer/end user faults?
- What is the service provisioning SLA per customer? For example, the requirement could state: 90% of all new orders must be fulfilled within three working days upon contract signature. This figure is measured on a calendar month basis.
- What is the fault response and fault resolution SLA per customer? For example, the requirement could state: 100% of all new faults reported by customers must be responded to and status updated within 1 h of a customer reporting the fault and 90% of category 1 faults must be fixed within 4 h of a customer reporting the fault. Are these figures measured on a per calendar month basis?

- What are the response times for service cancellation or termination? Although this will not have any SLA implications on the customers, this will have a huge implication on potential unused, nonbilling network resources. This is probably an internal SLA for efficient use of network resources.
- What is the service availability for the service? Is there a distinction between service availability and network availability for the particular service?
- What is the network availability for the service? What are the boundaries for the network availability? Is the service responsible for connections to a third party/OLO network?

3.2.4 Service Key Performance Indicators and Other Service-Related Questions

- If the service is going to be substituting an existing service (e.g. migrate from PSTN to VoIP service), what customer/end-user migration requirements are there? Is temporary loss of service during the migration acceptable to customers/end users?
- If the customers/end users are to convert from one service to another (e.g. converting from an integrated service digital network (ISDN) service to ADSL service), what are the customer/end-user conversion requirements? Is temporary loss of service during the migration acceptable to customers/end users?
- What are the business and operational key performance indicators (KPIs) for the service? This may include things like:

  o the revenue generated from the service in a year;
  o the average revenue generated per customer/end user;
  o the number of new customers or end users per month;
  o the number of customers/end users leaving the service per month;
  o the percentage of service provisioning within the customer SLA;
  o the average time, longest and the shortest times, it takes before a customer service agent answers a call when a customer/end user rings the service help line.

Further KPI reports example can be found in Chapter 8.

3.3 Market or Marketing Requirements

What are market requirements? These are requirements obtained from market analysis and customer feedback. For example, due to the growth in the number of applications being developed and the bandwidth requirements for each application on the network, there is a market requirement for the broadband access service to support different bandwidths.

In order to define market requirements, you need to understand the market you are trying to serve and define the sector and audience for your service. You cannot have a service that is universal for everyone, because people are different by nature and everyone has different requirements. What you can do is to find a common theme in the requirements and compartmentalize what is required.

Market requirements can also come from competitor analysis. For example, the market for a particular service will only have a launch window of the next 6 months, otherwise competitors will be launching a similar service after that. To be the market leader, the service must be launched within this window.

In addition, the marketing requirements may impose volumetric (nonfunctional) requirements (e.g. the subscriber growth, type and mix of traffic projected) on the service (which may be in different geographic areas) and potential systems requirements. For example, the marketing department may decide to run some promotional activities for a period of time in a local area. This promotion may lead to a sudden surge in volume of traffic in certain parts of the network (hence, to additional capacity requirements on the network), as well as to a surge in the number of customers/ end users ordering the service (hence, additional requirements for the system to support a higher number of simultaneous sessions and processing ability and additional operational staff to deal with the high number of orders). Additionally, if the promotion is related to price or discounts, then there will be requirements in the billing systems to ensure that the appropriate discount schemes and rates can be applied to the customers/end users and that they are billed accurately.

As you can appreciate from the above examples, market and marketing requirements have a wide impact on all parts of the service, from network and system capacity to operational requirements. Therefore, it is important to define, understand and analyze these requirements.

3.4 Reporting Requirements

Reporting requirements are often forgotten until the end. However, from the customers' point of view, this is how they can tangibly measure the performance of the service. From the business point of view, it is having these management reports that help to manage the service. Therefore, the two main sources of reporting requirements are internal business performance reporting and measurements of the service performance for customers. Reporting requirements for customer-facing reports are generally derived from the SLAs of the service.

Internal reporting requirements can broadly divided into the following areas:

- internal KPIs reporting (mainly service-related reports);
- order management performance reporting;
- fault management performance reporting (including performance reports of third-party suppliers and measurements of maintenance agreements with external agents);
- customers'/end users' service usage reporting;
- network incidents reporting;
- network utilization and performance reporting;
- capacity/resource management reports;
- accounting and revenue reporting.

Therefore, when capturing the reporting requirements for the service, it will be useful to talk to the people responsible for all the areas listed above. Here are some questions you should consider asking:

- What are the formats for the reports? Are the reports tabulated? Are graphic representations required? Or are both forms of presentation required?
- Will trend analysis be required? If so, what period are the trends to be charted over? Will it be month by month or week by week?
- What is the medium by which these reports are presented? Are they paper reports? Are they electronic reports?
- For electronic reports, what file types are required for all the reports?
- What are the confidentiality levels of the reports?
- Are actual figures required? If so, what format do they need to be in (e.g. in Excel or in a commaseparated file)?
- For customer reports, are the customers allow to download the actual data for electronic reports?
- Are all the reports read only? What write protections are required?

- What are the dispatch or distribution mechanisms for the reports? Can they be downloaded from a secure Web portal or FTP to a certain place or e-mailed to a pre-defined e-mail box? What security measures are required for delivery of reports?
- Who and where should the reports to be delivered to?
- Can internal users define their own reports based on the data available? If so, who will have the authority to approve those reports?
- What are the frequencies for each of the reports? Is it daily, weekly, monthly or on demand? It will be useful to define a day when the reports are going to be available. For example, monthly reports are on a per calendar month basis, available on the second of each month; daily reports are measured from midnight to midnight and will be available at 1 a.m. every day; hourly reports are measured on the hour and available 10 min past each hour.
- Do any of the reports need to be a real-time snapshot of the current state of play?

Further reporting requirements and questions to capture reporting requirements can be found in Section 8.14.

3.5 Security Requirements

The objective of maintaining security is to ensure that the right person (i.e. authentication of the person) with the right authorization has the correct level of access to certain services or data (both end-user personal data and network data). With the right security policy implemented, it also protects personal data and network data from being misused. From a service design point of view, some of the security requirements will be service specific and be derived from the service description (e.g. the service is to support secure socket layer (SSL)), but security requirements often arise from the protection from misuse of network and system resources and prevention of potential damage caused by hackers (who can be from within the organization or external). For end users of the service, privacy and confidentiality of their personal data, voice and data transmitted when using the service and their location will be of concern. This is more pertinent for mobile services, as the mobile access network is the most vulnerable to attacks. The security requirements should be defined in line with the security policy of the company; and they are not specific for a service, but general considerations for the operational environment.

With the above in mind, some fundamental questions you might want to ask when designing a service with regard to security follow below. These questions are for triggering your thoughts only and not meant to be exhaustive.

3.5.1 End-User Security

- For the particular service, who is responsible for authenticating the end user? What authentication protocols need to be supported? How much end-user verification is required?
- What are the end-users' confidentiality requirements? Can their personal details be used by a third party?
- Is there a prescribed mechanism or protocol required to protect the integrity of the communication channel (e.g. the support of SSL, hypertext transport protocol over SSL (HTTPS) encryption, cryptography or public key infrastructure (PKI) technologies)?
- What would be the mechanism required to ensure the integrity of the data being transported and that it is not corrupted?
- What would be the mechanism required to prevent session 'hijacking'? What session layer security (or security protocol) is to be supported?
- What privacy policy does the service need to adhere to? How do we ensure the end-user profile is not accessed by unauthorized personnel?

- What requirements does the data protection legislation have on the service?
- Is there a requirement to protect the end user from denial-of-service attacks?
- If the service has a feature for dynamic update of user profiles and service packages to be bought, what authentication requirements are there to ensure the users are authentic? Since changing service profiles may involve billing the end users for additional services, disputing the request for change will be costly.
- Is there a difference in requirements between upgrading and downgrading of end-user service packages?
- How do you protect the confidentiality of the voice and data being transmitted?
- For a mobile service, how do you protecting an end user's location privacy?
- Is application security a requirement? If so, what security mechanism is required? For 3G mobile services, the IP multimedia subsystem (IMS) is a special case for application-layer security.

3.5.2 Service Provider/Operator Security

- Does the service need to adhere to BS 7799/ISO 17799 or other security standards?
- How do you ensure data integrity within the network?
- How do you provide confidentiality protection for communication between different network equipment/elements?
- How do you control the physical security of the network equipment/elements, especially those that are outdoors?
- Is there any requirement for setting firewalls to filter out unwanted messages and potential attacks? If so, where is the most appropriate place?
- How do you control the logical and physical access of all network equipment/elements?
- What protection mechanism is required to ensure the integrity of signaling and control messages within the network?
- For a mobile service, how do you ensure the radio spectrum within your access network is not 'stolen' for malicious intent?

3.6 Functional Requirements

Functional requirements are very much dependent on the service to be designed. Most of the technical requirements are derived from the service description. However, there are some that are applied across all services. In this section, I attempt to ask questions that will help you to tease out the necessary technical and functional requirements.

3.6.1 Functional Requirements Questions

- What are the applications that the service is supporting? For example, for a broadband service, is there a requirement for voice or video applications? Does the service support broadcasting? Will the traffic be unidirectional or bidirectional? What video format is required? Will asymmetric upload and download speeds affect the service? Will there be a requirement for symmetrical upload and download speeds?
- Where are the boundaries for the service? For unmanaged services, where would the responsibility of the service end?
- What are the session-layer protocols to be supported?
- What are the transport-layer protocols to be supported?
- What file transfer protocols are to be supported?
- What signaling protocols are to be supported?

- What are the network interface specifications to be supported?
- What network diagnostic protocol is to be supported?
- Is mobility support required? If so, what geographic coverage does the service need to include?
- What are the external network interconnection specifications to be supported?
- Does this service need to be technically integrated with other services?
- How much system support do operational departments need? Will the existing system functionalities be enough for the service? Will you need to introduce new systems?
- Is there a particular system platform that needs to be used for any of the systems within the service?
- Is there a particular technology/technology platform that needs to be used for the service?
- Are there any system interface requirements for interconnecting internal and external systems together?
- What are the interface and graphical user interface (GUI) requirements for customer/end-user facing systems? What applications/services/tasks do these interfaces need to perform? (For example, if electronic bill is part of the service, what end user/customer interfaces are required?)
- What technical interface requirements/specifications do these external interfaces need to conform to?
- Is the service subdivided into different service packages for end users/customers to choose from? Can they be dynamically changed by the end users/customers or are they static and require manual operation to change?
- What are the upload and download line speeds to be supported by the service?
- Is bandwidth contention allowed for the service? If so, what is the maximum effective contention ratio that can be applied?
- Is there a restriction on how much content (data volume) a customer/end user can download per month? If so, what is the limit per service package? Once the limit has been reached, will the customer/end user be 'suspended' for the service or will additional charges be applied to the end user/customer?
- What network management protocols are to be supported?
- Are there any requirements imposed on the network from the operating systems and vice versa?
- What are the GUI requirements for all the internal systems that require human interactions?

3.6.2 Billing Requirements

In addition to the business requirements related to pricing and billing (stated in Section 3.2), below are some more technical questions to be asked related to billing. Most of them have technical implications with respect to data collection and billing system set-up to support the billing functions.

- Are different charges applied to different applications? For example, is voice charged differently from data applications or video applications?
- Are the one-off set-up charges for each application used? Is the set-up charge per session, or is it on per end user/customer connection basis?
- Are different charges applied to different QoS levels? Or is this accounted for by using different service packages?
- If calls are delivered to other network providers (OLOs, e.g. mobile to fixed line calls), will they be charged differently from those delivered within the operator's own network? Do different OLOs have different rates?
- For mobile services, what is the charging structure for a roaming call between operators and calls roamed outside of the home network?

- Are there any intercharging requirements for packet-based traffic being exchanged between network operators? If so, what are the charging structures (who pays for which leg of each call/session)? Is it on a per session basis or on a time duration basis or the amount of data exchanged?

3.7 Network Planning Requirements

Network planning requirements are necessary to ensure the network is planned and built in accordance with the service requirements. These normally come from the following areas:

- the forecast of number of end users/customers for the service;
- the forecast of the type of traffic/application to be used and bandwidth requirements per end user/customer for the service;
- the marketing/special promotion plans for the service, which could affect end user and network traffic volume;
- the type of user equipment (UE)/CPE projections;
- the projected geographic coverage of the service;
- the lead time required building the network capacity required.

For example, it is anticipated that the demand for the fixed broadband will grow by 10% a year and that 100% of the population want the service. From this, a geographic roll out of the service will require planning the access and core network capacity in order to cope with the demand forecasted. Also, the scalability of the solution will also be looked at in order that this demand can be fulfilled.

Most of the network planning requirements are deduced from the market size and the forecast provided by the product management area. Typically, network planning requirements also take into account the population distribution by geographic area (for end-user services), the location of businesses/business parks (for business services) and the location of motorways and major train lines (for mobile service). The service designers need to ensure that these network nonfunctional requirements listed above are captured for the network planners to perform capacity and network planning analysis.

3.8 Nonfunctional Requirements

Nonfunctional requirements are often forgotten, but these requirements are essential to make the customers/end users feel that they are getting a good service and to make the service work. It is absolutely no use if there is a lot on functionality in a service but the interface is so user unfriendly that it becomes unusable and the end users avoid using the service or use the service provided by the competitor instead. Or perhaps your service has lots of features, but the customers need to wait ages to get the service provisioned or for your systems to respond. Therefore, nonfunctional requirements specify how the service is delivered to the customers/end users and not what is being delivered (those are functional requirements). Without nonfunctional requirements, your service will be useless! Quite often, it is the nonfunctional requirements that distinguish your service from your competitors, especially now that all the functional features between service providers are becoming very similar. Taking response time as an example, if your service has a slower response time to customer requests (e.g. session set-up or call set-up), then the perception is that your service is of poor quality and customers will look for alternative suppliers.

Writing precise and testable nonfunctional requirements is often very difficult. There are many factors and parameters to consider and it is difficult to define numerical measures in a meaningful manner. Taking the response time example again, if you specify a response time of 0.5 s for session set-up time for a broadband end-user session, does it have to be 100% of the time? What load consideration is there for that response time? What is an acceptable lower limit? One way to get around this problem is to state the target figure and also the expected figure to be reached a percentage of the time (e.g. 80%).

Trying to achieve certain nonfunctional requirements (e.g. performance requirements) may also be costly and the cost can easily outweigh the benefits. With the response time example, the cost of 0.5 s may cost $10 per end user, while a 1 s response time may cost $2 per end user. Therefore, when defining nonfunctional requirements, it is worth bearing in mind the potential cost. Unrealistic nonfunctional requirements should also be avoided.

Nonfunctional requirements are also very difficult to measure and test. Setting up a test environment for testing nonfunctional requirements can be costly. Hence, it is important to define nonfunctional requirements up front to ensure the cost of testing is captured as part of the business case for developing the service.

Nonfunctional requirements are divided into the following broad areas:

- **Performance.** From a service point of view, this is mainly referring to response times of systems or call/session set-up time under a certain load; the speed at which the services are provided; the amount of degradation of service under load and resource utilization of network and systems. These requirements specify need for the entire end-to-end service, as this is what the customers/end users will experience. It is not a lot of use to specify a response time for a particular system unless the function only involves one system. Specifying meaningful service performance requirements is not that easy. The performance requirements should also includes the behavior of the service when the service is close to under its maximum load.
- **Availability.** This is the percentage of time when the service is available (i.e. when the service can be used). The service availability could include both network and system availabilities. Planned outages are not normally included in the availability figure.
- **Reliability.** This defines how often systems or network equipment can break. This will affect the availability of the service. Often, the mean time between failures (MTBF) figures are specified for this type of requirement.
- **Robustness and error rate.** This defines how easily the service can be broken and how often error occurs. This is often linked to the loading of the service (e.g. what is the system load before the end users/customers will experience a degradation of service?) Again, the availability of the service is affected by the ability to fulfill the robustness requirements.
- **Usability.** These requirements refer to the human interactions between the different functions/features supported by the service and the end users/customers. These interactions could be through the network terminal equipment (e.g. the mobile telephone) or via a portal (e.g. access to service performance reports via the Internet). These requirements can be translated into GUI requirements. Usability requirements can be defined by using the following parameters: the speed at which user can learn to use the service; what documentation and training materials are required for customers/end users; online help requirements and standards to be adhering to.
- **Throughput.** This is rate of data flowing through the network or systems. Throughput can also be part of the performance of requirements, but, in the context of telecommunication services, throughput requirements are probably best listed separately, as the service is complex and different functions and features may vary. The throughput requirements determine the processing power requirements for the equipment of the network and systems. Bottlenecks which may limit the service throughput should be avoided when designing the service.
- **Maintainability and upgradeability.** These requirements define the ease of maintaining and upgrading the service. They should define if service downtime is acceptable when upgrading

the network or systems and if backward compatibility on different versions of software on customers'/end users' terminal equipment is a requirement. If downtime is acceptable, then the acceptable length for downtime and the definition of acceptable downtime window should be specified (e.g. planned downtime can take place between 1 and 5 a.m. on weekdays and each outage window should be less than 2 h). If backward compatibility is required, then you need to specify how many versions behind the current one will be supported by the service.

- **Recovery and data integrity.** These requirements specify the ease of recovery and the data integrity requirements in the event of systems/equipment failure. This also includes back-up requirements. Mean time to recover (MTTR) figures are often used to specify these requirements.
- **Error and exception handling.** These requirements define how the service behaves under error conditions and the maximum impact the end users/customers should have under error conditions. Definition of how exceptions are handled with the service should also be stated within this requirements area.

Security is sometimes classed as a nonfunctional requirement. Below are some questions that may help to arouse your curiosity on nonfunctional requirements.

3.8.1 Nonfunctional Requirements Questions

- What is the response time for each feature and function of the service? What are the load conditions for these response times and with what percentage of the time within what time period?
- What is the throughput for each feature and function of the service? What are the load conditions for these throughputs and with what percentage of the time within what time period?
- What is the session/call set-up time requirement and under what load conditions? What is the highest acceptable limit and under what load conditions?
- If there is a malfunction in either the network or system supporting service, what is the acceptable impact on the service and what is the time required for the alarm for this malfunction to be triggered?
- What is the maximum amount of simultaneous alarms/alarm triggers that the network management system will need to handle for this service? One probably needs to define 'handle' within the context of the service being offered, as 'handle' is not testable!
- For each of the features or functions of the service, what is the maximum time required for the end users/customers to learn to use the service or features or functions within the service?
- What are the requirements for end users'/customers' training materials, instruction manuals or documentation required for the service?
- Will online help be a requirement for the service? Or is a technical help line required?
- What are the user interface requirements for the service? A special icon required at the users'/customers' terminal equipment? Will there be a special application to be loaded onto the terminal equipment?
- How will the service be initiated from the end users'/customers' point of view? Pressing a button? Starting up an application?
- When initiating the service or a feature/function of a service, what mechanisms do the end users/customers need to go through before the service or service session or feature/function is set up or activated?
- When initiating the service or a feature/function of a service, how many steps or keystrokes do the end users/customers need to go through before the service or service session is set up or activated?

- Is it acceptable if the service or 'live' customer/end user session is affected if one of the processing cards of one of the network elements or service systems is pulled out?
- Will it be acceptable to introduce service downtime when upgrading network elements/service systems? If so, what are the maximum time and the outage window?

3.8.2 Network and System Utilization and Performance Requirements Questions

- What are the network and system utilizations on an hourly basis on specified network links and systems respectively?
- What are the network and system utilizations on a monthly basis on specified network links and systems over a time period of 3 months? There may be a requirement to review utilization on a quarterly or on a per 6 months basis.
- What are the port utilization requirements for the switch network (this could be applied to both PS and CS networks).

3.8.3 QoS Requirements

- Are there any QoS guarantees to be supported (e.g. minimum throughput rate of x)?
- What type of traffic shaping policy can be applied?
- Is QoS a feature of the service? If so, what are the QoS parameters to be supported? For example, network latency to be <10 ms 90% of the time or round-trip delays <5 ms 95% of the time, packet loss to be $<y$ per day, call set-up time of <1 s 100% of all calls, call or session teardown time to be <0.25 s 100% of all calls/sessions.

3.9 Regulatory, Licensing and Legislation Considerations

As a result of the privatization and liberation of the telecommunications industry (from the Telecommunications Act 1984 in the United Kingdom, the Telecommunications Act 1996 in the United States and subsequently the WTO General Agreement on Trade in Services (GATS) in February 1997), regulations in different countries have been set up by the governments of those countries to provide a framework for fair competition. Privatization of fixed telecommunications in the United States, United Kingdom and European Union simulate competition, and regulations and licensing become an important means to administer the pro-competition market environment. Different countries have different regulatory and legislation requirements. From a service point of view, the service you are launching may benefit, but must adhere to the relevant regulations, legislations and licensing agreements; otherwise, your company will face immense consequences. Below are some examples that illustrate how these factors affect different services.

3.9.1 Local Loop Unbundling

Local loop unbundling (LLU) is the process where the incumbent operator is to make their local access network (the copper cables that run from customers' premises to the telephone exchange) available to other companies. This enables alternative operators (as opposed to the incumbent) to upgrade individual lines (using digital subscriber loop (DSL) technology) to offer a broadband service directly to their customers.

Many developed countries, for example the United Kingdom, Australia and member states of the European Union, have introduced regulatory frameworks for LLU. LLU is regulated in two main areas: the process by which the alternative operators can obtain access to the local access network and the pricing of the LLU offerings by the incumbent.

3.9.2 Mobile, 3G Licensing and Spectrum Ownership

3G licensing and spectrum ownership was the subject of intense debate in the early 2000s, where bidding for 3G licenses and spectrum was a fashionable (but also expensive) thing to do. However, obtaining the license also came with conditions you need to fulfill. Some examples are given below.

- **Coverage and rollout obligation.** Under most of the 3G licenses that have been issued, the licensees have to roll out the network to cover a certain geographic area or a certain percentage of the population within a defined timeframe. This will give the operator a big headache, as rolling out a mobile access network is expensive!
- **Spectrum allocation.** Most frequency licenses are not transferable. Therefore, unused spectrum cannot be traded between operators. This is imposed to avoid abuse in the spectrum market. Therefore, the licensee has to use the allocated spectrum for their services.

3.9.3 Data Protection Act 1998 (United Kingdom)

The principle of the Data Protection Act 1998 in the United Kingdom is to protect personal data from misuse. All operators and service providers (in the United Kingdom) need to adhere to this act. The text below has been quoted from the Ofcom (Office of Communications) website (http://www.ofcom.org.uk/about/cad/dps/eight/):

In order to respect people's privacy, service providers need to process personal data in accordance with the data protection principles, namely:

1. Personal data shall be processed fairly and lawfully.
2. Personal data shall be obtained only for one or more specified and lawful purposes, and shall not be further processed in any manner incompatible with that purpose or those purposes.
3. Personal data shall be adequate, relevant and not excessive in relation to the purpose or purposes for which they are processed.
4. Personal data shall be accurate and, where necessary, kept up to date.
5. Personal data processed for any purpose or purposes shall not be kept for longer than is necessary for that purpose or those purposes.
6. Personal data shall be processed in accordance with the rights of data subjects under this Act.
7. Appropriate technical and organizational measures shall be taken against unauthorized or unlawful processing of personal data and against accidental loss or destruction of, or damage to, personal data.
8. Personal data shall not be transferred to a country or territory outside the European Economic Area unless that country or territory ensures an adequate level of protection for the rights and freedoms of data subjects in relation to the processing of personal data.

Therefore, the above provisions have to be made when operating a service (in the United Kingdom).

3.10 Financial Constraints

Financial support is never infinite. Therefore, financial constraints are a major consideration when a new service is being considered. The amount you are allowed to spend when developing a service is dictated by the figures that have been approved in the business case. A viable business case will be accompanied by a financial model. The financial model estimates the revenue, operating cost, capital expenditure, general administration cost and taxes. Although you might have an approved business case, you may not be allowed to spend all the cash within one financial year, as your company's cash flow may prevent you from doing so. Therefore, the service may be launched in phases as a result.

In addition, the network cost model and analysis is not often done to allocate costs to the various services using the network. The operational cost of each service is not often accounted for. Hence, the accuracy of calculating the profit or loss of a service is often impaired.

Financial constraints do not always impose negative effects. For example, financial constraints may result in more efficient use of network and system resources. From a network point of view, this is where capacity planning and traffic management for the service (or all the services) comes into play.

3.11 Physical Location and Space of Network Equipment and Systems

Physical locations and space are normally seen as constraints on a service. The physical locations of network elements and systems are normally dictated by the concentration of ingress and egress traffic sources and destination locations. Sometimes, spaces for equipment can be hard to find and it may be more economical to use the transmission network to transport traffic to a location where space is available rather than leasing another building just for one service.

On the access network front, network equipment is normally housed in the exchange building where possible. Planning permission will be required for establishing masts for mobile services, and this may have implications on the physical location of the base transceiver station. Planning permission might take some time and will certainly have timescale implications on your project plans.

3.12 Service Description Template: A Service Description that Fully Defines the Service

This section provides a summary/template of items to be included in the service description. The items listed cannot be exhaustive, as each service and company is different, but this should provide you with a good checklist to verify your service description.

3.12.1 Purpose and Scope

The purpose of the service description is to describe the service from the customers'/end users' point of view. This document describes all the service features, functions and business drives and so on of the service. However, the description on how the service works, in terms of network and systems, should not be described here, but references should be made to the technical service description (please see Chapter 5 for details) or other design documents. The service description should remain as an internal document. Marketing literature is the medium for describing the service for external use.

3.12.2 Target Markets, Benefits and Value

This section describes the target market for the service. Any market segments that are excluded should be stated with reasons for exclusion. The size and value of the target market for the service are shown here. Benefits to the target markets are also listed.

3.12.3 Channels of Sales

States the intended channels to market for which the service is developed.

3.12.4 Benefits to the Customers

Benefits of the service to the customers/end users are stated here.

3.12.5 Benefits to the Resellers

Benefits of the service to the resellers (if relevant) are stated here. Commission processes and criteria to be met by the resellers are also listed here.

3.12.6 Service Launch Timetable/Roll-Out Plans

The service launch timetable and the roll-out plans for general availability of the service should be stated here. Any phased launch dates are also highlighted, stating the features available/geographic coverage for each phase.

3.12.7 Description of Service

This section provides an overview of the service. Items to include are (but not limited to): why the company is offering the service; how the service is used; which geographic area it will serve; types of application to be supported; and so on.

3.12.7.1 Service Features

This section should list all the service features and the combination of features or service packages allowed for the service. Sometimes, network features are also listed. Any service bundles with other services within the company's portfolio and their combination should also be highlighted in this section. Any available service variants and service variant combinations are stated here, together with any unavailable options or certain feature combinations that are not technically possible or very expensive to implement.

For example, the service features may include:

- Bandwidth requirements for end users/customers. It is useful to specify upload and download speeds, as the bandwidth for both may be different.
- Network protocol supported.
- Applications scenarios to be supported (voice, data, video, video on demand, e-mail or the combination of applications to be supported). This should include a description of how the customers/end users can use the service and the different areas of applications.
- Managed network service with defined boundaries (e.g. managed virtual private network (VPN) up to the egress Ethernet port of CPE).
- Web hosting service (e.g. with x Gbit storage, managed application service, managed database service).
- For voice/VoIP services, the features may include call forwarding, voice mail, three-way conference calling, call blocking, call waiting, call transfer, different ring tones, caller ID and so on.

- Security features or managed security service (including firewall management, intrusion detection, antivirus management, authentication and authorization, PKI certificate management, etc.).
- Types of CPE and network interfaces supported, including the recommended CPE under different bandwidth/application scenarios. The conditions under which the equipment is kept and the specification of the equipment should be listed. Definition of how and for whom the CPE is installed should be stated.
- Types of customer interface and how the customer can access the service. If the service is to be provided by a third party, then please state the boundaries of responsibilities of the service. This is best illustrated in a diagram.
- If resellers are involved as part of the service, then interfaces to the resellers are described.

3.12.7.2 Future Enhancements

Future enhancements and their timescales of availability should be highlighted here. Any features and geographic coverage that are not planned (with reasons) are stated here.

3.12.7.3 Service Network Architecture

A high-level network diagram of the service may be useful, but this will only be finalized when the service has been designed. This section of the service description should not be written until all the design activities have been completed. Otherwise, you will be imposing a solution from the service description.

3.12.7.4 Service Components

These can refer to different service features that are allowed and the service components available. For example, the service components may include:

- a list of network terminal equipment/managed CPE provided;
- different service profile and associated network features;
- access network bandwidth;
- different grades of service levels (e.g. gold/silver/bronze level of service);
- list of customer reports;
- grades of security packages.

These service components are normally tied to billing items, as each service component may be chargeable or priced differently.

3.12.7.5 Billing and Charging

The billing mechanism and pricing schemes of all the service components are described in this section. For example, a one-off installation and a recurring rental will be charged for each managed CPE in the solution. The installation charge will be charged on the month of installation and the recurring rental will be charged quarterly in advance. The CPE rental charge will be pro-rated if installed in between billing cycles.

If the service has usage-based charging, then the charging scheme needs to be described in this section. For example, calls between 08:00 and 18:00 will be charged a/min during weekdays and b/min will be charged at other times and public holidays.

Any discount, service bundling discounts or revenue share scheme of the service should also be included in this section. Details on amount or discount percentage, the criteria and approval process for discounts and revenue share scheme are to be detailed here.

Payment terms, billing accuracy, bill format, billing support information (e.g. call records), medium of bills, billing support information distribution and bill payment methods of the service should also be described here. Minimum contract terms and penalty charges are also specified.

If the service is sold through resellers, then commission or discounts to be applied upon reaching a certain sales volume should also be stated here.

3.12.7.6 Service Performance

This section describes how the service performance is measured. This includes all the service KPIs for the service as well as the customers' SLA and their associated penalties. Please see Section 3.12.11 for customer SLA template.

3.12.7.7 Customer Reporting

A list of reports available to customers should be listed here. Most of the customer reports are related to SLAs (please see Section 3.12.11 for details). The mechanism of delivery should also be stated (e.g. is it over a Web portal or printed paper copies or e-mail to customers). It is important to define the customer reports up front, as it has huge implications on data collection during the service design phase.

3.12.8 Customer-Facing Support Functions and Process Arrangements

To achieve the desired customer experience, it is important that the service description specifies how various customers-facing functions are expected to handle various service requests. The main areas are:

- **Customer engagement**. The channels, target customer and contact management processes are stated here. During the customer engagement process, the amount and type of sales support required should be highlighted. If customer network or solution design activities are required, it will be useful to state the skills of the personnel required. Sales and commissioning process (for both direct and resellers) and system requirements are also described.
- **Order handling**. This describes how the orders will arrive and what customer interfaces are available for order placements and order status checking. The different types of order allowed (e.g. moves, termination) and the criteria/conditions/minimum terms to be met are stated here. It will be useful to state how service termination, cancellation, moves, additions and changes are handled and through which interfaces.
- **Fault handling/management**. This highlights the fault reporting and escalation interfaces customers have access to. The mechanism of providing fault status updates should also be stated.
- **Customer help desk functions**. It will be useful to list all the customer help desk functions. For example service enquiry, fault escalation, billing enquiries, customer complaint handling, customer technical support, etc. The operational support hours are stated here.
- **Service maintenance**. This includes processes for carrying out scheduled and emergency service maintenance activities. Disruptions to the service should be minimized for all service maintenance activities. The process should include the assessment of impact to customers/

end users for each scheduled service maintenance activity and ways to minimize disruptions should be described. If emergency service maintenance is required, then the process should describe the criteria for emergency maintenance and how this can be carried out to minimize service impact.

- **Billing and revenue assurance**. The billing mechanism and billing processes are described here. It should include items like billing fulfillment processes, bill payment methods, bad debtors definition and processes for dealing with customers not paying their bills (e.g. service termination due to nonpayment). Other revenue protection and assurance processes should be described here (e.g. credit control and bill reconciliation processes).
- **Financial management**. Any processes regarding the financial management of the service should be described here. This may include processes to manage cost and gross margin, financial outlay to operate the service, service credit issuance and revenue reporting processes.

The requirements for customer support functions are stated here. However, the solution and the operational support processes for the customer service will be written after all the service design activities have been completed.

3.12.9 Support System Requirements

Support systems (both operational and business support) requirements can be stipulated from the service description and the customer-facing support in Sections 3.12.7 and 3.12.8. However, one might like to state these requirements explicitly in a separate section. This can be part of the service description or in a separate service requirement document. Areas to be covered include:

- **Fault monitoring and management**. This describes the need of preventive fault detection and pre-emptive measures for fault scenarios. The requirements for fault management systems are listed here.
- **Order management.** This describes the system requirements for handling and fulfilling customer orders.
- **Service and SLA management.** This describes the requirements for service management and SLA management. This includes the items to be measured in order to ensure that the customer SLA, in terms of service availability, customer provisioning, fault management target and so on, is met and that the internal service KPIs are measured. Any internal reports to manage the service (e.g. network efficiency and occupancy, system utilization, customer satisfaction) are also listed here.
- **Forecast, capacity planning and traffic management**. This describes the forecasting process, roles and responsibilities for producing and managing the sales forecast and the assumptions and methods for producing forecasts for the service. This may include a summary of the sales forecast and operational volume metrics. Triggers for capacity upgrades (both network and systems) are listed and the items to be managed in order to avoid traffic congestion on the service are listed. Key financial items in terms of network and system capacity upgrades are also identified.
- **Fraud, revenue assurance and financial reporting.** The system requirements to ensure revenue for the service are collected and any potential fraud scenarios for the service are stated here. Reports for measuring the financial performance (e.g. revenue of the service) of the service should be listed.
- **Billing.** All the billing and tariff requirements for the service are described here. This includes the billing format and structure required for the service, all the billing elements to be charged and their associated tariffs, and any discounts schemes and structure to be applied.

3.12.10 SLAs

SLA is a mechanism by which the customers measure how well your service is performing. It should include a list of objective measures to demonstrate the service performance. Normally, service credit payouts are associated with each SLA measurement. Below are some example service-level measurements you can consider:

- service provisioning lead time;
- response time for a fault reported by customer;
- guarantee time to fix a customer-reported fault after the receipt of customer fault report;
- network change request lead times;
- service change request (i.e. upgrading or downgrading service features/packages) lead times;
- customer-facing systems response times for each request;
- service availability – you may consider separating service availability into network availability and customer-facing systems availability;
- billing accuracy, punctuality of bills deliveries and handling of customer disputes;
- time required to deal with billing enquiries;
- time required to resolve customer complaints;
- call or session set-up time;
- network latency;
- packet loss;
- QoS measures for different quality of service levels (e.g. gold, silver or bronze service). Each service level may have a different definition to the following parameters: packet loss, network latency, peak information rate, burst size, bandwidth, queue depth and so on.

3.12.11 SLA Definition Template

Table 3.1 gives some sample SLA definitions for you to consider.

Further SLA definition examples and definition recommendation can be found in *SLA Management Handbook Volume 2 – Concept and Principles* [55] and *SLA Management Handbook Volume 3 – Services and Technology Examples* [56].

3.12.12 Marketing Support and Collateral

All the marketing collaterals and literature to help customers using the service are listed here. These documents are issued to external parties. They may include:

- marketing collateral for sales and marketing;
- marketing collateral for resellers;
- commissioning schemes and qualification criteria for resellers;
- support interfaces for resellers or business-to-business (B2B) customers portals.

Any presales support requirements (e.g. presales resources and training requirements) are also stated here.

3.12.13 Contract/Terms and Conditions

Describe the terms and conditions of the service that are not already covered as part of the general contract terms of sales for the company. For example, the minimum term for service provision, the condition by which the customers/end users can terminate the service and any condition that

Table 3.1 Sample SLA definitions.

SLA measures	Target service levels
Service provisioning lead time	Customer network provisioning – 60 days upon contract signature End user provisioning – 5 working days, 90% of the time, measured on a per calendar month basis
Response time for a fault reported by customer	Initial response within 1 h upon receiving customer fault report. Subsequent updates will be every 2 h
Customer-reported faults resolution times (upon the receipt of customer reports)	Category 1 faults resolved within 4 h Category 2 faults fixed within 8 h Category 3 faults resolved within 12 h Category 4 faults resolved within 24 h Definition of each category can be defined as follows: Category 1 – total loss of service Category 2 – partial loss of service Category 3 – suffering from service degradation Category 4 – cosmetic
Network change requests lead times	30 days depending on network capacity availability
Service change requests lead times	1 day for both upgrading and down-grading of service features/packages. Service downgrade is subject to min. contract terms and conditions
Customer facing systems response times for each request	Authentication of user login – 3 s for both successful and failure attempts On-line bill viewing – bill available for viewing within 3 s upon request Online bill download response time <2 s upon request (excluding time required to down load bill) Viewing of customer reports – each report is available for viewing within 3 s upon request Downloading of customer reports – download response time <2 s upon request (excluding time required to download) Display of fault report updates – within 3 s upon request Display of order status – within 3 s upon request
Service availability	Network availability of 99.9% Customer facing system availability of 99.8%
Billing accuracy, punctuality of bills delivery and handling of customer billing disputes	Billing accuracy 99% Punctuality of bills delivery 99% Handling of customer disputes – response within 2 days upon receipt of complaint. Resolve 90% of bill disputes with 1 month
Time required to deal with billing enquiries	Average time to respond to billing enquiries is 1 day over a monthly period
Time required to resolve customer complaints	Average time to resolve customer complaints is 3 days over a monthly period
Call or session set-up and tear down time	Call set-up time of <1 s 95% of all calls and <2 s for 100% of calls Session set-up time of <1 s 95% of all sessions and <1.5 s for 100% of sessions Call or session tear down time to be <0.25 s 100% of all calls/sessions
Network latency	Network latency to be <10 ms 90% of the time; <100 ms 100% of the time
Packet loss	Packet loss to be <100 packets per day

include guarantees levels and QoS. Any special contractual/legal framework between the company and the customers/end users are identified here.

3.12.14 Legal, Regulatory and Interconnect

Any legal and regulatory requirements to be adhered to for the service or for the customers/end users to use the service are described here. The operator license and interconnect agreements relating to the service are stated. Any specific legal or regulatory obligations (e.g. universal service provision or provision of emergency service access) are listed. Providing a list and cross-reference to any regulatory requirements for the service or for the customers/end users to use the service may be useful.

3.12.15 Technical Constraints on the Service

All the technical constraints for the service (e.g. service access options, geographic coverage, network capacity) should be listed here. This should be written after all the service design activities have been completed.

3.12.16 Training Requirements

Training requirements for internal and external staff should be stated here. Internal personnel requiring training may include sales, presales support, marketing sales support, customer technical support and all internal operational support areas and so on.

External personnel training requirements should also be listed, for example resellers training requirements.

3.12.17 Related Documents

A list of related documents should be stated here. This should include:

- sales forecast;
- service roadmap and roll out plan;
- technical service description of the service;
- network design documents;
- IT system design documents;
- operational processes for the service.

3.12.18 Glossary of Terms

It is useful to have a list of definition of terms to ensure that there is no misuse and misunderstanding.

3.13 The Success Criteria for a Service

Having defined your service, how do you measure that it is successful? You may say that the KPI defined for the service should indicate the success of the service, shouldn't it? Well, KPI is definitely a good indicator for the health of the service (otherwise you need to redefine your KPIs), but it does not necessarily measure the success of the service.

Measuring the success of the service depends on the objective(s) of the service you have defined. If the objective of the service is to acquire and grow your market share, then measuring for growth and growth rates will need to be defined. If the objective of the new service is to be the market leader, then market penetration will be a good indicator. Therefore, the success criteria of the service depend very much on the objective of the service. Below are some indicators that may help you to measure the success of the service.

3.13.1 Growth in Market Share

Measuring the growth rate in the market share is a good indicator for measuring the success of the service against your competitor. Growing your addressable market share is great. However, one needs to be aware that having strong growth does not always translate to greater profitability for the company. Only quality growth will enhance your profitability. In addition, you may achieve that growth at the expense of other areas of the business or as a result of a very expensive marketing campaign, neither of which delivers healthy profit.

3.13.2 Market Penetration

Market penetration is a great success indicator if you are first in the market with the service. Like measuring the growth of market share, it should not be done at the expense of other areas of the business while achieving the desired market penetration. Only good market penetration with quality customers will generate healthy profit.

3.13.3 Churn Rate

Churn rate can be a useful measure for the success of the service if you are trying to retain current customers and improve the perception of the current service. Managing the churn rate of the service is becoming more important, as the service features offered by different service providers have little differentiation between them and the pricing is becoming competitive. In addition, it is more expensive to gain new customers than to retain current customers. Therefore, the churn rate should be kept to a minimum.

Your target churn rate could be 5% over a period of 6 months. You could be running loyalty promotional programs to retain current customers. Each promotional program should have a metric to track its success rate.

3.13.4 Revenue from the Service

With a sizable market share and a low churn rate, the revenue of the service must be great, right? Not necessarily! Your service might be attracting customers who have a bad credit history and do not pay their bill. You may also want to compartmentalize your revenue streams to measure and analyze the success rate of the service by different sale channels. For example, revenue (and percentage of revenue) generated from direct sales may be less than that of resellers. Then, your sales strategy may need revising. Therefore, keeping track of the revenue of the service will also reduce the risk of revenue leakage within the service and analyzing the source of revenue will help to identify potential sales opportunities.

3.13.5 Customer Satisfaction Survey

Churn rate is a typical measure, and probably the first indicator, for customer satisfaction. The churn rate will be low if the customers/end users are happy. However, if the service is new and there is a minimum term of contract associated with it, then the churn rate might be low within the contract term, but once the contract is up for renewal, the churn rate might go through the roof if the customers are not happy with the service. Therefore, performing a customer satisfactory survey will help the company to gauge the level of satisfaction amongst your customers. Providing a proactive customer care service will also help to improve the customers' perception of the service.

3.13.6 Planned Service Outage Reports

If the service has many hours of planned service outage, then by default the customers cannot be getting a good service. Although the service availability figure is looking relatively healthy (as the service availability figure normally excludes planned outages), you are still having many technical problems if the number of hours of planned outage is high. The planned service outage report should tell you how much actual downtime the service is experiencing and, hence, the health of the service.

4

Service Design Process

In any design and development book, the development process is one that we cannot do without. It is virtually impossible to run any business effectively and deliver high-performing services without the appropriate management processes in place. The uniqueness for service design is that it touches all parts of the company. Therefore, trying to design a service without a company-wide management process is doomed to failure.

On the face of it, the management and development process for service design is no different from that of any other design or software development process. However, what I am trying to achieve in this chapter is to link the service design process with the business approval process at various development phases. This is designed to minimize risks and to be efficient with the project expenditure without jeopardizing the project timescale. This is especially important for telecommunications developments, as they involve significant investments.

Also included in this chapter are:

- the organizational changes and structure required for introducing new services;
- potential resource requirements;
- the program management structure and roles and responsibilities of different stake holders;
- documentation and control structure for successful service design.

For those who are familiar with the Enhanced Telecom Operations Map (eTOM) [31], the service design process described in this chapter is an expansion to the Service Development Process (process identifiers 1.2.2.3 and 1.2.3.3) and Product & Offer Development & Retirement process (process identifier 1.2.1.5). The processes described in eTom cover the service design and development area at a very high level. This chapter defines, in more detail, the steps required to design, develop and implement a service solution. It also provides the linkages to the business approval process.

This chapter describes the service design process, whilst Chapter 5 describes what needs to be done at each stage of the process in detail.

Successful Service Design for Telecommunications   Sauming Pang
© 2009 John Wiley & Sons, Ltd

4.1 What are the Key Steps to Develop New Services and Service Enhancements?

The service design process illustrated in Figure 4.1 is in the purest form. It is very similar to any product design or development process. Figure 4.2 includes all the documentation to be produced in each phase.

4.1.1 Phase 1: Concept

Service concepts and ideas can be generated from anyone and from anywhere within the business. Sometimes it is technology driven; other times, it is customer or market driven. Therefore, the concept stage is designed to throw out any hair-brain ideas that are impossible to implement or to operate. This phase is also used for any customer special requests for services that are not part of the service portfolio (for customer special requests/developments, please see Section 4.5 for details). At the end of the concept phase, the sales and marketing teams will know if the requests are technically possible together with rough cost and implementation timescale.

The depth of the analysis and time to be spent in this phase should be kept to the minimum as at this stage, the ideas may or may not be feasible. It is also much more cost efficient to decide in the early stages of the Service Design process if the ideas or concepts are worth pursuing.

There will be scenarios where the designers require further information or requirements to proceed. It is at this phase where requirements are clarified.

4.1.1.1 Phase 1 Documentation

At the end of phase 1, the service designer should have completed the rapid impact analysis, stating if the ideas/concepts can be implemented, together with rough estimates of lead time and cost. Please refer to Section 4.7.1 for the content included in the rapid impact analysis document.

4.1.2 Phase 2: Feasibility

From the concept phase, we should already know if the ideas are technically feasible. Only concepts that make business sense (i.e. are commercially viable) will go through to the feasibility phase.

The main purpose of the feasibility phase is to identify the potential solutions and the cost of each solution option. In this phase, more detailed requirements are also identified. Quite often, not all internal requirements are recognized from the concept phase.

As part of the feasibility study, proof-of-concept testing may also be carried out on potential solutions. This will help with the selection of the eventual solution. Most of the time, proof-of-concept testing is on the technology performance. It is a good idea to involve technical operational personnel with the testing so that they can evaluate the operability of the potential solutions.

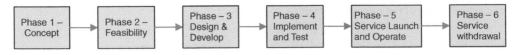

Figure 4.1   Service design process.

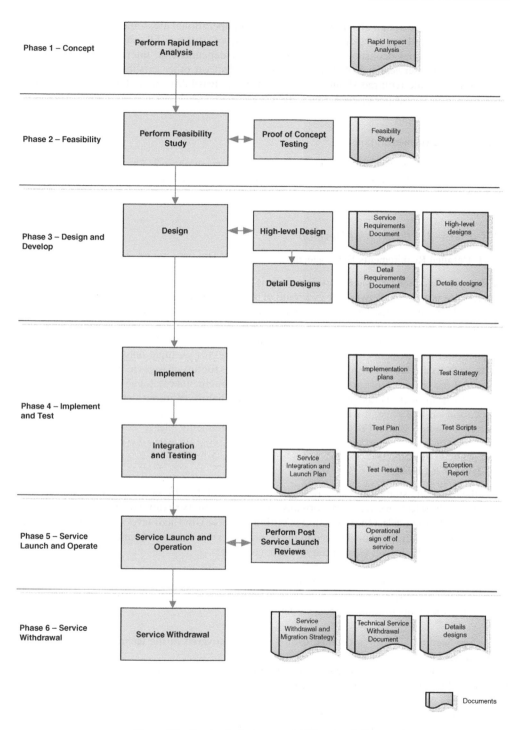

Figure 4.2 Service design process and documentation.

As with all service design activities, the impact on all operational areas and IT systems must be assessed. The cost of any development activities (e.g. new features in IT systems, additional head counts in operational department, space constraints), dependencies, interservice dependencies between different services (i.e. if this service is dependent on other existing services), potential issues and timescale required should be captured in the feasibility stage. Factors influencing a feasibility study are further discussed in Section 5.2.

4.1.2.1 Phase 2 Documentation

The signed off feasibility study is the main output document in phase 2. This document should be signed off by representatives of all areas impacted. The content of a feasibility study can be found in Sections 4.7.2 and 5.2.

4.1.3 Phase 3: Design and Develop

4.1.3.1 Detailed Requirements Development

The first activity in the design and develop phase is capturing a set of detailed requirements for the service, if this is not already available as a result of phase 2. Most of the high-level requirements will be derived from the service description and they need to be translated into detail requirements and technical specifications for various parts of the solution. These are service requirements. For example, the network features to be offered will need to be translated to network functional and nonfunctional requirements. Any systems-related feature, for example billing requirements, will need to be translated to billing requirements/specifications for any changes to be made to the billing system. All the requirements are logged (ideally in a requirements catalogue/database) and prioritized. This is an onerous task and should not be taken lightly. Without detail requirements and specifications at this stage, it will be impossible to verify the design and plan the testing activities required in the next phase. For details on gathering requirements, please see Section 5.3.

4.1.3.2 Finalized Design

From the feasibility study, there may be design options available. In the design phase, the solution is finalized. The selection criteria are normally based on the ability to fulfill the requirements stated, the cost and the timescale. Once the solution option is selected, the high-level solution of the service is formed. This solution is documented in the technical service solution document. This document should provide a high-level view of the solution, covering:

- the network topology of the solution;
- the high-level system architecture and/or system impact for all business and OSSs;
- the high-level operational process model/impact.

Please see Section 4.7.4 for a detailed listing of topics covered in the technical service solution documents.

Detailed designs for various parts of the solution are also developed once the high-level design is finalized. This could be an iterative process, where feedback from the detail design might influence the high-level design.

It is almost unavoidable that some initial laboratory testing and prototyping may occur before the final design can be completed. It is important, therefore, to allocate enough time, resources and budget to this activity when planning this phase.

4.1.3.3 Prepare for Implementation and Testing

The implementation plan and test strategy can also be developed once the solution has been confirmed. These should be available before proceeding to the next phase. This will give the business an idea of timescale so that sales and marketing departments can start finding lead/trial customers. For telecom services, it may be very difficult to have a prototype for the whole service due to the excessive resource requirements. A technology prototype may be relatively easy to set up. This can be part of the feasibility and design process. However, running a trial service can be costly. It is not unusual to run a market trial or have a trial service in a certain geographic area before roll out to a wider audience. However, I will class this type of trial as an operational service. Although the scale of operation is smaller than the eventual service, you still will have paying customers/end users using the service.

4.1.3.4 Phase 3 Documentation

A set of design documentation will be produced in the design phase. They are:

- service requirements document;
- detail requirements documents for network, IT systems and operational processes;
- technical service solution document;
- network high-level design document;
- IT solution/business and OSS high-level design document;
- high-level operational process document;
- detail design documents for all the elements in the solution.

The structure of these documents will be discussed in Section 4.7.3.

4.1.4 Phase 4: Implement and Test

4.1.4.1 Implementation

When all the designs are finalized and resources have been agreed, the service implementation can go full steam ahead. As well as devising the implementation project plan, you might now want to consider your implementation strategy and devise how the service can be implemented most effectively. Do you want to build everything yourselves? How much do you get your contractor to do? These are some fundamental questions you need to answer before the implementation plan can be drawn up. Implementation strategy is discussed further in Chapter 10.

4.1.4.2 Technical Integration and Testing

The integration and test phase ensures that all the components of the service integrate and work seamlessly together. Very often, many projects fail because they underestimate the resource and expertise required to perform this task. Integration and testing requires specialist resources and a significant amount of time to complete. How many projects have you worked on that have squeezed testing till the last minute and failed? Any testing methodology book will tell you that the verification process starts from requirements and ends when the system is operational, in this case service launch. You need to start planning the testing activities as early as possible. It is also a good principle to have a separate team dealing with testing rather than having design teams verifying and testing their own designs. This will enable independent design verification and, consequently, design flaws can be spotted earlier in the design phase.

Integration and testing activities may include:

- unit testing of individual network elements and IT support systems;
- technical integration testing of various network elements and IT support systems working together;
- service integration testing – ensuring that all the service built is fit for purpose and no element of the service is missing;
- operational testing – ensuring all the operational processes defined are appropriate and that the operational staff are trained and available.

Details of the testing required are covered in Section 5.5 and Chapter 11. Testing and its methodology is a subject of its own. In this book, I only extract the high-level principles and activities that are relevant to service design. Useful references for software testing include: *Software Testing and Continuous Quality Improvement* [22] and *Software Testing and Analysis: Process, Principles and Techniques* [25]:

4.1.4.3 Service Integration Strategy

Service integration strategy is similar to that of a technical integration test strategy. The purpose of the service integration is to ensure the interactions between other services are well integrated and that the service works end to end, including the operational processes for the service. Often, the technical integration and test strategy is part of the overall service integration strategy. Different factors for service launch and integration will be further described in Chapter 11.

4.1.4.4 Phase 4 Documentation

Documents to be produce in this phase include:

- implementation strategy and plan;
- implementation project plan;
- technical integration and test strategy;
- test plan;
- test cases and test scripts;
- test results;
- list of exceptions;
- service integration and launch plan.

Topics to be covered within all the above documents can be found in Section 4.7.6.

4.1.5 Phase 5: Service Launch and Operate

This is the phase where the service is in live operation. It is inevitable that there will some teething problems at the start of the service. Therefore, having a lead customer or trial customers at the start of the service can provide the operational teams with the opportunity to run the service on a smaller scale and to iron out any problems before rolling out the service to more customers/end users. It is unavoidable that some of the operational processes or some support systems will need small modifications as a result. It is good idea to hold regular post-service launch reviews and maintain the design team in place to resolve any issues that may arise. The post-service launch reviews also provide a channel for customer feedback and to collect requirements for future service enhancements.

4.1.5.1 Phase 5 Documentation

The main document to be produced in this phase is the operational sign-off for this service. Other documents may include the post-service launch review notes and actions to improve the service.

4.1.6 Phase 6: Service Withdrawal

A service may come to its natural end when either the technology has become obsolete or when the service is on longer in demand. Withdrawing a service is as complicated as introducing new ones. After scoping the service withdrawal activities, a withdrawal strategy is devised detailing the approach of the withdrawal, the risks involved and the resources (labor and costs) required. This should also include the migration strategy of customers from the current service to another replacement service. Service withdrawal is detailed in Chapter 12.

4.1.6.1 Phase 6 Documentation

The documents for service withdrawal are:

- service withdrawal strategy document;
- technical service withdrawal design document or high-level technical redesign documents;
- detailed redesign documents.

Sign-off from various operational departments is required, as they will be most affected by this activity. Further details on the above documents are covered in Chapter 12.

4.2 How Should the Process Link to Business Approvals at Various Phases of the Development?

After looking at the service development in its purest form, it is time to put it in to practice. It is fundamental to have the service design process tie in with the business approvals process. This is illustrated in Figure 4.3. I am also assuming that the business approval process sets the direction for service sectors or customers the company wants to invest in.

4.2.1 Gate Review

Gate review is a forum consisting of representatives from all parts of the company impacted by the introduction of new services. Typically, this forum has representatives from the business units (e.g. product management, sales – if it is a customer special request), finance department (if it has a major financial impact or impact on financial systems), design departments (e.g. service design, network design, IT, program management) and operational departments (e.g. customer service department, service provisioning department, network operations, first- and second-line support for faults, billing operations). In the gate review meetings, representatives of each department can decide if the project can proceed to the next phase. Quite often, resources of each development are assigned in this review. Reasons for not proceeding are also raised in this forum. This is a mechanism to manage business risks for each business area.

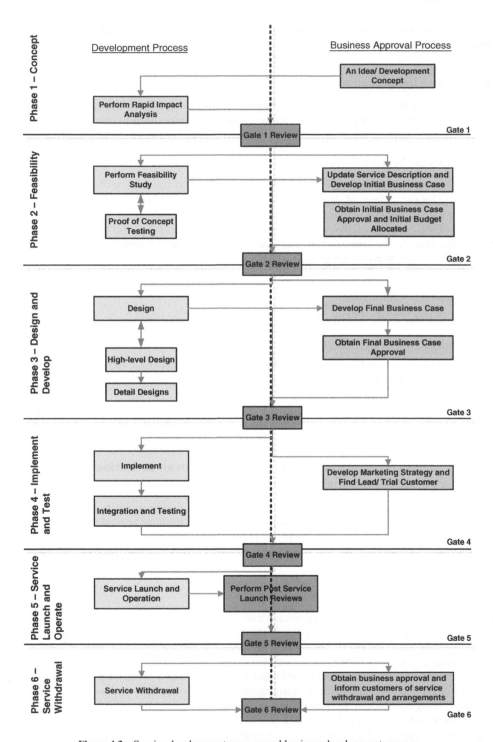

Figure 4.3   Service development process and business development process.

4.2.2 Phase 1: Concept

As described in Section 4.1.1, service concepts and ideas can be generated from anyone and anywhere within the business. The persons raising the concept requests will detail their ideas in the first part of the rapid impact analysis for effective evaluation. This is the mechanism used to communicate to the design team, and the rest of the business, the service concept description, the market potential, competitor position, time window for this new service opportunity and high-level requirements. Please see Section 4.7.1 for the rapid impact analysis document structure.

Once resource is allocated for this analysis, the service designers perform the requirements analysis, technical evaluation of potential solutions and talk to various operational departments for initial impact analysis. As this analysis is meant to be rapid, time spent on this phase should not be more than 2–3 weeks. If more time is required, then a feasibility study should be conducted. The gate review forum can agree that the business should take the service concept to the feasibility stage without a completed impact analysis.

4.2.2.1 Output

Output of phase 1 is the completed rapid impact analysis. The completed analysis should contain potential solutions, rough estimated costs and any clarification required before proceeding to the next phase of the development.

4.2.2.2 Gate 1 Review Result

Based upon the completed rapid impact analysis, the business can decide if the service concept is fit for the next development phase. The outcome of the gate review meeting can be:

- approve to go through to next phase;
- no go;
- on hold.

The reasons for each outcome should be recorded.

4.2.3 Phase 2: Feasibility

Phase 2 of the development is performing the feasibility study. As stated in Section 4.1.2, the feasibility study should contain potential solutions and costings for each solution. It should also include the pros and cons for each option.

During the feasibility phase, requirements are clarified and are fed back to the product management. This will enable them to refine the service description, emphasizing which functions and features are a priority for implementation.

The costs of all the options should be listed. Ideally, in the feasibility phase, the cost of testing should also be included. Most of the time we only account for the cost of implementation and forget about testing until too late. It is important that all costs are identified, as these figures are used in the initial business case for this new service concept. Estimated costs should have a percentage variance on the accuracy attached so that, when using the figures for the business case, the business can understand the potential risks of this cost variance.

Since the costs and impact will be used in the initial business case, it is important that the feasibility study has obtained the sign-offs from the various impacted departments before costs are used in the business case.

4.2.3.1 Output

Outputs of phase 2 are:

- signed off feasibility study;
- updated service description;
- approved initial business case and initial budget allocation.

If the initial business case is not approved, then the result of the gate 2 review meeting is automatically a 'no go'. No gate review will take place.

4.2.3.2 Gate 2 Review Result

Based upon the completed feasibility study, the business can decide if the service concept is fit for the next development phase. The outcome of the phase review meeting can be: approve to go through to next phase; no go; on hold. The reasons for each outcome should be recorded. After the gate 2 review approval, an initial budget is allocated to the design phase. This will limit the risk and financial exposure for the business.

4.2.4 Phase 3: Design and Develop

The cost of developing and operating the service will be finalized after the solution has been selected. This will be apparent when all the high-level designs are completed. The cost items should include all network elements, IT system developments and upgrades, all implementation cost, resource allocation for development (design and testing) and ongoing operational cost, cost of testing (equipment and test laboratory, etc.).

Since this costing will go into the final business case of the service, it is important that the costing is as accurate as possible. The final business case is the final green (or red) light for the development of the service from a business financial support point of view. If the final business case has not been approved, then the development of the service will stop and no gate 3 review will be required.

4.2.4.1 Gate 3 Review Result

The result of the gate 3 review can be to approve to proceed to the next phase or to put on hold.

You may ask, 'How can the project be on hold if the final business case been signed off?' The answer is simple: lack of resource is one of the most common occurrences for putting projects on hold.

4.2.5 Phase 4: Implement and Test

Having an appropriate implementation strategy is the key to the success of this phase, as selecting the right approach will potentially save a lot of time, resources and money. This is a nontrivial subject, as it involves much supplier management and project skills to have everything implemented within timescale and budget. Please see Chapter 10 for further thoughts.

Testing is an activity that everyone knows we need to do – make sure it works, right? I think it is more than that. Apart from making sure the service is fit for purpose (i.e. fulfilling the requirements) and that it works as designed, the fundamental principle of testing is to minimize the risks (financial, business and technical) when a new service is introduced. For example, the last thing you want to see is for the new technology for the new service to jeopardize the integrity of the current network being operated or for the new support system to require 10 extra, unbudgeted, people to operate.

4.2.5.1 Gate 4 Review Result

After the gate 4 review, the result can be to approve service launch or to reject service launch.

There are many reasons for rejecting service launch. One of the most common is too many test failures or many things need to be fixed before launching the service. This also means that the risk of launching the service is too great for the business. In this scenario, all the problems need to be fixed (or design revised) and all the relevant tests need to be rerun and another gate review needs to take place.

4.2.6 Phase 5: Service Launch and Operate

It is suggested that post-service launch reviews are to be held regularly for the first 3–6 months of new service operations. This forum will facilitate the resolution of cross-departmental issues that may arise during the initial service launch, as well as obtaining customer feedback. Further improvements to the service can be also made through this forum.

It is also good practice to have a few trial customers before roll out to more customers (general availability). This should give all operational areas a chance to sort out any teething problems before further customers are added to the service. In this scenario, having post-service launch reviews will be vital.

4.2.6.1 Gate 5 Review Result

Gate 5 review only happens when the service is at the end of its useful life. The result of the review can either be to approve or reject service withdrawal. A draft of the service withdrawal proposal is to be presented at the gate 5 review.

4.2.7 Phase 6 – Service Withdrawal

Before the service can be withdrawn, product management needs to assess the benefit of withdrawing the service and what communication will need to be sent out to the customers and internal departments. If the service has come to its natural end and no customer is left on the service, then the withdrawal process is a lot simpler. Other business considerations may also include how to reassign the operational personnel that are currently operating the service. Do they become redundant? Do they have to be retrained to operate other service? Upon the business approval of service withdrawal and the sign-off of the final withdrawal strategy, budget will be allocated and withdrawal activities can take place. Please refer to Chapter 12 for further details on service withdrawal.

4.2.7.1 Gate 6 Review Result

The result of the gate 6 review could either be service withdrawal completed or failed. This gate review should be held at the end of the service withdrawal unless some major issues arise that will hinder or stop the service being withdrawn. The gate review forum can decide to abandon the service withdrawal and mitigate any risk that is associated with it. This is equivalent to gate 3 of the service withdrawal process defined in Chapter 12.

4.2.8 Summary

Table 4.1 provides a summary of events within the service design process. It also highlights the stakeholders at different phases within the process. Parties to be consulted and informed can vary between organizations; the list is long and, hence, not stated here. However, this is covered in more detail in Chapter 5.

Table 4.1 Summary of events in the service design process.

	Phase 1 – Concept	Phase 2 – Feasibility	Phase 3 – Design & Develop	Phase 4 – Implement & Test	Phase 5 – Service Launch & Operate	Phase 6 – Service Withdrawal
Input	Service concepts with description of service concept and market analysis	– Business approval to enter feasibility Phase. – Definition of service and detail market analysis.	– Initial Business case approval – Service definition finalised – Resource for Phase 3 from all departments required allocated	– Final business case approved and budget allocated – All design documents signed off – Resource for Phase 3 from all departments required allocated	All exceptions from the testing phase have mitigation plans and work around.	– Service withdrawal strategy ready – Project plan for implementation completed – Resources required allocated
Output	Completed Rapid Impact Analysis with rough costs	Feasibility Study with design options and estimated cost for each option	All aspect of the design finalised and signed off. Final cost of the whole development identified.	– All aspect of the service implemented. – All test cases complete and test results available. – All exceptions listed	– Operational sign off. – Post service launch reviews notes and actions to improve the service to customers	– All customer/end users migrated off the service – All equipment and network resources released. – Re-assign operational resources.
Gate review input criteria	Completed Rapid Impact Analysis with rough costs	Completed and signed-off feasibility study.	– Signed off high-level and detail design documents.	– All operational processes defined and operational staff is available and trained.	– Justification for withdrawing the service.	– Service withdrawal status.

Gate review result	Approve to go through to next Phase/no go/on hold	Approve to go through to next Phase/no go/on hold	– Final business case approved – Implementation plan and Test strategy available. Approve to go through to next Phase/on hold	– Service integration plan signed off. Approval/reject of service launch	– Agreed service withdrawal strategy. Approval or rejection for service withdrawal.	– Exceptions listed. Service withdrawal complete/fail
Stake holders (Responsible, Accountable, Consult and Inform – RACI)						
Responsible	Product management	Service design	Service design	Programme management	Head of operation	Programme management/Head of operation/Product management
Accountable	Service design	Service design	Design authorities of various network, systems and operational areas	Design authorities of various network, systems and operational areas for implementation. Integration and test manager for testing	Head of various operation departments	Head of various operation departments
Consult						
Inform						

4.3 Organizational Changes and Structure Required to Develop New Services

4.3.1 Operational Resources

Trying to implement changes in the operational environment is not easy. Most of the time, operational personnel are far too busy operating (managing network events or customer change requests, etc.) and do not have time to deal with new services to be introduced. They can also be resistant to change as a result of the new service. Therefore, having a change agent/manager within each of the operational areas is essential to successful new service introductions to the operational environment.

Ideally, the change agents have the operational experience within their respective areas and are able to perform analysis on the new service requirements, with the skills to design new/additional changes to the existing operational processes within those areas. They also need to have a good knowledge of the support system being used, as they are required to help with defining changes in the support systems for the service. If a new system is required to support the new service, then the change agent will help to define the user requirements (sometimes, the functional and nonfunctional requirements) for the new system and will also be responsible for defining the operational acceptance/user acceptance testing (UAT).

Since all operational processes need to work together to provide an end-to-end service, it will be wise to have a person (process design authority) to coordinate all operational areas and run process workshops where necessary to ensure all required operational processes are defined and that no gaps exist between the processes. Ideally, this person should also have operational experience and can identify the process needs. The process design authority is also responsible for working with the change agents to define the operational training plans/courses for each operational area.

4.3.2 System Analysis Resources

From the support systems point of view, business/system analysts are assigned to the project to evaluate the new service requirements and translate these service requirements into system specifications for system developments. Like operational processes, all the systems need to work together to provide an end-to-end service support. Therefore, the IT system architect of the company or the system design authority should perform such a role.

4.3.3 Network Technical Resources

For the network technologies, network designers and a network design authority are required to design the service network and decide how best to configure the network for this service. They are also responsible for selecting the network technologies for the service (if required) and design how best to fit the new technologies into the existing network environment.

4.3.4 Service Design Resources

From past experience, virtual teams work very well when designing services. The virtual team should consist of representatives/change agents from all parts of the business and should be formed in phase 3 of the service design process; however, during the feasibility phase (phase 2), the service designer will be talking to all parts of the business to complete the feasibility study, and network design resources may also be required to perform a technical network feasibility on potential network solutions. The service designers lead the virtual team and have the authority on all aspects of the service from a design point of view. The service designer is also the main interface to the project/program management and internal customers (e.g. product managers and sales). An example of the virtual team structure for service design activities is as below (Figure 4.4).

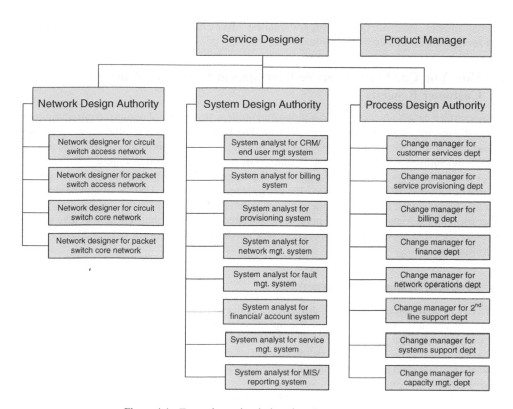

Figure 4.4 Example service design virtual team structure.

4.4 Resource Requirement for Designing Services

It is useful to have estimations of the resource requirements for designing and implementing a new service, especially at the gate reviews, where resource allocations take place. This, of course, varies between different companies. Table 4.2 provides an example matrix on resource profiles for

Table 4.2 Example of resource requirements for designing a service.

Resource requirements	Classification of service		
	Type 1	Type 2	Type 3
Service designer	0.5	1	1.5–2
Process design authority	0.5	1	1
Network design authority	1	1	1
IT systems design authority	0.5	1	1
Network designer	0	1	2–3
IT systems analyst	1	2	3–4
Program manager	0.5	1	1
Network implementation project managers	0.5	1	2–3
IT system implementation project managers	1	2	3
Operational change managers	3	4	5–6
Service integration manager	0.5	0.5	1
Estimated development lead time	2–3 months	6 months	6–9 months

different project sizes, with type 1 being the smallest. Over time, the resource profile can be refined with more accurate resource estimations.

4.5 How You Can Use the Service Development Process for One-Off Customer Requests

From a service design point of view, designing and implementing special requests from customers is essentially another service development activity. The task may be of a smaller scale, but, nevertheless, the same development steps need to be followed, as the analysis process is the same and that operational support will need to be put in place. The process is detailed in Figure 4.5.

For smaller changes, the concept and feasibility stage can be combined, but business approval will still be required before proceeding to the design stage. The approval process might be simplified with one business case approval for the customer solution. Depending on the scale of the solution, integration and testing activity may need to be adjusted. It is likely that testing will take place at customer sites that are currently in operation. One needs to be aware of the risks and how they can be minimized.

4.6 Program Management Structure

Different companies are organized differently. However, the program delivery team for new services is normally a virtual team, made up of representatives from all parts of the business. The program manager is ultimately responsible for the delivery of the development and owns the development budget as assigned from the sign-off of the business case. Figure 4.6 is an example program management structure. I have identified the key stakeholders for delivering new service developments. You can adapt this to your own organization.

Program/project management is a subject on its own, and I do not intend to cover it here. Some useful references include:

- *The Project Manager* (FT Prentice Hall) by Richard Newton [4];
- *Project Management: The Managerial Process* by Clifford F. Gray and Erik W. Larson [37].

4.6.1 Roles and Responsibilities of Different Stakeholders

Table 4.3 provides some examples of program stakeholder roles and responsibilities.

4.7 Documentation and Control Structure for Service Design

As stated in Figure 4.2, below is the list (not exhaustive) of documents produced as part of the service design process:

- rapid impact analysis;
- feasibility study;
- requirements documents;
- service design documents;
- implementation strategy and plan;
- test strategy, test specification.

Figure 4.7 shows the interrelationships between the documents. Content descriptions for these documents are detailed below.

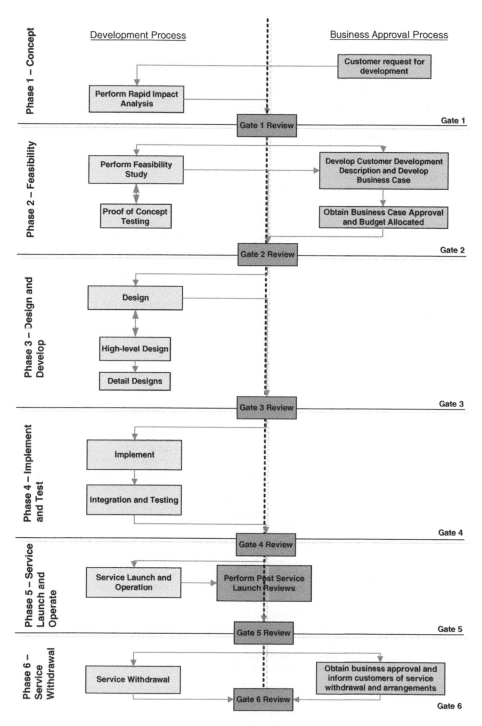

Figure 4.5 Customer special service requests development process.

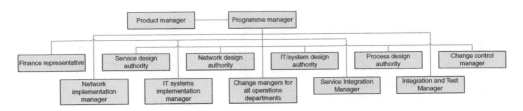

Figure 4.6   Example program structure for service development program.

Table 4.3   Example program stakeholder roles and responsibilities.

Stakeholders	Roles and responsibilities
Program manager	Responsible for the delivery of the development Owns the development budget as assigned from the sign-off business case Communicate to senior management regarding the progress and issues of the development Resolve any issues and anticipate potential problems that hinder the progress of the development Manage risks as they arise and identify potential risk and find ways to mitigate them Manage the progress of the development program Control the gate reviews
Product manager	Act as the project customer for service development Owns the service from a business point of view and has the ultimate say on service direction Define service definition Identify lead/trial customers
Service designer/service design authority	Act as the design authority for the service Responsible for all design aspects of the service Analyze service requirements and develop a solution for the service Produce rapid impact analysis, feasibility study, technical service solution document, service requirements document and detail requirements documents (where appropriate) Access impact on the service for any new feature/change request during the development Provide major input into gate review
Network designer/network design authority	Act as the design authority for the service network Responsible for all aspects of the service network Analyze network requirements and develop a network solution for the service Produce high-level and detail network design documents and detail network requirements document (if appropriate) Access impact on the service network for any new feature/change request during the development Provide major input into gate review
IT/system design authority	Act as the design authority for the support system (business and operational systems) for the service Responsible for all aspect of the support systems for the service Analyze systems requirements and develop the systems solution for the service

Table 4.3 (*continued*).

Stakeholders	Roles and responsibilities
	Produce high-level and detail systems design document and detail systems requirements document (if appropriate) Access impact on the support systems for any new feature/change request during the development Provide major input into gate review
Process designer/process design authority	Act as the design authority for the operational processes for the service Responsible for all aspects of the operational processes for the service Analyze process requirements and develop a process solution for the service Work with all operational change managers to produce high-level process design and detail work instructions for each operational department Ensure all necessary detail work instructions are in place for all operational departments supporting the service Produce detail process requirements document (if appropriate) Access impact on operational processes for any new feature/change request during the development Define training plans for all operational areas for the service Provide major input into gate review
Network implementation manager	Implementing the network design as detailed in the detail network design document. Project manage all network implementation activities Manage the network implementation budget
IT systems implementation manager	Implementing the IT/systems designs as detailed in the detail IT/systems design documents Project manage all systems implementation activities Manage the systems implementation budget
Change managers for all operations departments	Perform analysis on the new service/process requirements Design new/additional or change to existing operational process in the respective areas Help to define changes required in the IT/systems to support the service or network management function to be performed in the respective area Help to define the user requirements (functional and non-functional requirements if relevant) if a new system is required
Finance representative	Assess any impact on the finance department Ensure the program is within budget Resolve any financial issues the program might have
Integration and test manager	Ensure all the different components of the service (i.e. the service solution) work together seamlessly Manage all technical integration and testing activities (including requirements validation, design verification and testing activities) Define technical integration and test strategy Ensure the test facilities/environment are built and can perform meaningful tests in the facilities Escalate any issues arises from the testing phase to program management Ensure the testing is complete and that the risk of exposure has been minimized Provide major input into the gate reviews in the later phases

(continued overleaf)

Table 4.3   (*continued*).

Stakeholders	Roles and responsibilities
Service integration manager	Ensure all the different service aspects (i.e. the interservice solution) work together seamlessly Manage all service integration and testing activities Define service integration and test strategy Ensure all operational testing has been performed Ensure all exceptions from the operational testing logged and mitigation actions are in place and process modification has been performed Ensure all operational personnel have been trained and are available for service launch Escalate any issues arises from the operational testing to program management
Change control manager	Define and manage the change control process within the program Document change requests Track status of all the change requests Obtain sign-off for all the change requests Assess the impact of change requests to the program Identify when the change can be/needs to be implemented Log all change requests and actions with regard to the change requests

4.7.1 Rapid Impact Analysis Document

Within the rapid impact analysis, the following topics should be covered:

- market overview and proposal (including market analysis and business justification);
- service requirements (including all service features);
- constraints (e.g. critical timescales – if the service is not launched within the next 6 months, the window of opportunity will be gone; therefore, essential solution components are to be implemented first and other nonessential components will require subsequent phasing);
- potential network solutions and impact on current network, systems and operations for each solution option;
- financials (rough estimates of OPEX and CAPEX);
- timescale estimates for design and implementation.

The first two topics/sections of the document should be completed by the person raising the request (mainly product managers or a member of the sales team). By completing the first section of the rapid impact analysis, the service designer and the rest of the gate review forum will have a much better appreciation of the service concept. This is also a good tool to help product managers and the sales team to think through the ideas. Further details on the output of rapid impact analysis can be found in Section 5.1.

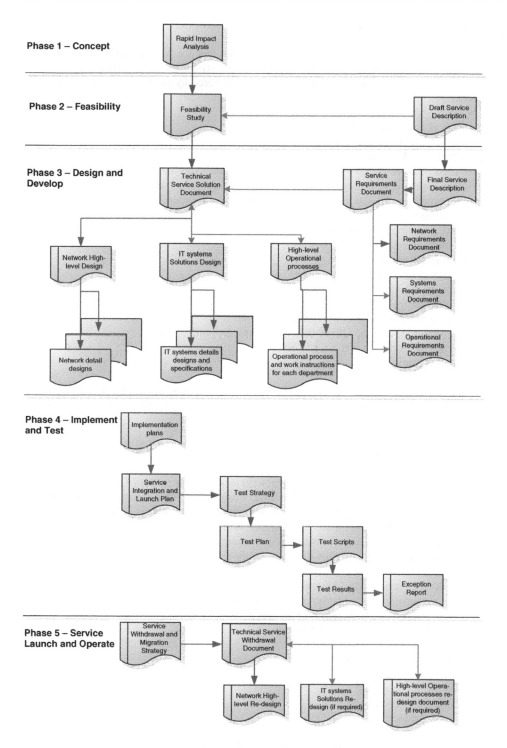

Figure 4.7 Service design documents structure.

4.7.2 Feasibility Study Document

The purpose of the feasibility document is to communicate to the business the solutions options for the new service and their associated implications, potential issues, risks and cost profile for each option. Some suggested topics to be covered by the feasibility study document are as follows:

- business requirements and drivers;
- service features and requirements;
- current network architecture;
- proposed network design options (stating the pros and cons and limitations of each option);
- proposed system solution options (stating the pros and cons and limitations of each option);
- impact on all operational areas (including processes and potential head-count implications);
- traffic/volume forecast;
- capacity planning implications;
- fraud implications;
- security implications;
- interconnect and regulatory dependencies;
- interservice dependencies;
- limitations (if any);
- risks, assumptions and potential issues;
- cost profile (CAPEX and OPEX breakdown and write-off costs);
- timescales (key milestones if any);
- recommendation/conclusion.

The feasibility study must be signed off by the relevant parties, as this has potential implications for the initial business case. Further detail on performing feasibility studies is discussed in Chapter 5.

4.7.3 Requirements Documents

Service requirements are mainly derived from the service description. Other service requirements may be imposed on the service due to existing operational constraints or from other external factors, like regulatory requirements or licensing requirements, where they may not be part of the service but be imposed by external issues. From the service requirements, the network, systems and operational process requirements will become apparent.

Requirement topics can be broadly divided into the following categories:

- Network-orientated service features and requirements;
- support systems-orientated features and requirements;
- nonfunctional and performance requirements;
- operational and system user requirements;
- pricing and commercial requirements;
- service management requirements;
- security requirements;
- regulatory requirements;
- third-party involvements;
- physical requirements.

More details on service requirements are covered in Chapter 5.

4.7.4 Design Documents

4.7.4.1 Technical Service Solution Document

The technical service solution document is an umbrella document that describes the solution of the end-to-end service as a whole. It describes the solution of the service at a high level and it is the only technical document that has a full view of the service. Where relevant, it references to the detail design documents for the different parts of the solution.

At the start of phase 3, the technical service solution document will drive all the other high-level designs. When the detail designs have been completed, the technical service solution document will be updated to include any additional details that might be needed or any changes as a result of the detailed designs.

Topics to be covered in the technical service solution document include all the topics that are discussed in Chapters 6–9. The detail content of the technical service solution document can be found in Section 5.3.3.

4.7.4.2 Network Design Documents

4.7.4.2.1 High-Level Network Design Document

The high-level network design document describes the service network architecture. It details what network components are required for the service solution and how the service will work from the network point of view.

4.7.4.2.2 Detail Design Documents

Detail network design documents take the high-level network design to a more detailed level. It describes the 'nuts and bolts' of the service network and explains how each of the network components is to be built and configured for the service to function.

4.7.4.3 Systems Design Documents

4.7.4.3.1 High-Level System Design Document

The high-level system design document describes the IT systems solution to support the service. It details how the various network and service management systems will support the service, the interactions between the various systems to form the system solution and describes how the various systems will interact with the network to form the service.

4.7.4.3.2 Detail System Specification/System Design Documents

Detail systems specification/system design documents specify system functions and the design of the individual systems within the systems solution. These design/design specification documents are used either to build the system or to specify changes required to the existing system. In most telecommunications companies, there is already a set of systems performing the service and network management functions. Therefore, introducing a new service will be a change to the existing systems. There are also cases where the new service requirements far exceed the current system capabilities, and new system development activities will take place. The system specification will be used to build the new system or assess potential 'off the shelf' solutions.

4.7.4.4 Process Design Documents

4.7.4.4.1 High-Level Process Design Document

The high-level process design document describes how the various operational processes work together to support the service. It details which operational department is performing which tasks.

4.7.4.4.2 Detail Operation Work Instructions

These are work instructions detailing the tasks the operational personnel need to perform for any manual operational functions for the service. These can also be detailed step-by-step instructions (e.g. perform task 1 then task 2 and so on) or detailed instructions on how to use the systems to perform certain operational tasks.

4.7.5 Implementation Strategy and Implementation Plan

During the implementation stage, one needs to decide how best the solution is being delivered. Do you use a third-party supplier to deliver the solution? Do you build everything in-house? Do you outsource the whole implementation to another company? The implementation strategy sets out how the implementation phase is to be carried out, and this is the basis where the implementation plan is developed.

The implementation strategy is formed based on the objectives of the project, as well as on what is being implemented. Detail on implementation strategies can be found in Chapter 10.

4.7.6 Test Strategy, Test Plan and Test Specification

Testing is a specialty on its own and I have no intention to cover it in detail here. However, below is an overview of the documentation expected from the testing phase and a brief description of the purpose of each document. Further details on service integration and testing are covered in Chapter 11.

4.7.6.1 Test Strategy

The main objective of a test strategy is to set out the approach to the integration and testing phase of the service design process. Within the testing phase, there are different types of testing activity taking place:

- unit/detail testing;
- system/network testing;
- network and system integration testing;
- end-to-end technical service testing;
- operational service testing.

Owing to timescale, the testing phase may quite often be subdivided into several stages/phases. The test strategy should state:

- the scope of the testing activities;
- how the different types and stages of testing fit together;
- the entry and exit criteria and output for each stage of testing;
- what is being tested in each type and stage of testing;
- the test team structure – role and responsibilities of the test team (e.g. who is responsible for writing which test specification and who is performing which tests);
- where (which test laboratory) and what testing facilities/environments are required to complete each of the test stages;
- documentation structure in various test phases (if there are more than one);
- test pass and failure criteria;

- fault and error reporting and escalation procedures;
- configuration management of test platform;
- post-testing activities (e.g. the network needs to be reconfigured or what test data need to be deleted before the service is launched, and so on).

Further details on service integration strategy can be found in Section 11.2.

4.7.6.2 Test Plan

The test plan is a high-level summary of tests to be performed. It should list all the areas to be tested and what tests will be performed in each area. This is useful for showing test coverage against requirements.

4.7.6.3 Test Specification

The test specification details all the test cases (including all the steps to be performed for each test) and test scripts to be performed, the expected results and the test pass/fail criteria.

Further details on integration and testing, the test plan and test specification can be found in Chapter 11.

5

Service Design: What Needs to be Done

Having understood the service design process discussed in Chapter 4 (illustrated in Figure 4.2), it is time to explore what actually needs to be done during the birth and the withdrawal of a service. This chapter describes and outlines the activities to be performed at each phase of the service design process as described in Section 4.1. Examples will be given to illustrate how various activities are performed.

5.1 Performing Rapid Impact Analysis

As described in Sections 4.1.1 and 4.2.1, performing a rapid impact analysis is meant to be rapid. It should not take more than 2–3 weeks for one service designer to perform the analysis. The objective of the rapid impact analysis is to assess whether the concept is viable and to ascertain the rough cost (±100%) of the development. If the analysis is taking longer than 2–3 weeks to complete, then the analysis is far too detailed and those details should be done as part of a feasibility study. The major differences between the rapid impact analysis and the feasibility study are the amount of investigation being carried out and the depth of the analysis. This should be limited to a 2–3 weeks' 'rough and ready' assessment, because the concept at this stage of the process has limited business approval and it could be a nonviable service. The input requirements at this stage are at a very high level. More details and information provided at this stage will increase the accuracy of the estimated development cost and timescale. More time spent on this analysis will mean higher business risk, as only minimal business approval has been sought at this stage.

Input into the rapid impact analysis should include (but is not limited to):

- a description of the service;
- market proposition of the service;
- business justification of the proposed service;
- market and competition analysis of the service;
- a list of service features (network, systems and operational requirements);
- constraints (e.g. critical timescales, nonoptional solution components and required phasing);
- service support/service management requirements;

- pricing constraints and commercial requirements;
- critical timescale (e.g. if the service or customer feature is not ready by a certain date, the deal will be over).

Taking into account the above input, the service designer will need to interview the relevant network technology experts, the system experts for the systems required to support the service and various operational departments to obtain their input to evaluate potential solutions and assess the impact in all business areas. As part of the analysis, service requirements will be clarified. If it is for a customer's special request, then requirements are to be clarified with the sales team.

The output of this analysis should include (but is not limited to):

- potential solutions and solution options;
- pros and cons of all the options;
- financial implications (CAPEX, OPEX, one-off costs and implications in the current and next financial years);
- third-party involvement (if any);
- timescale (lead times and any key milestones/date to meet the required time constraints);
- recommendations.

5.2 Performing Feasibility Studies

The objective of performing a feasibility study is to investigate:

- whether the service requirements can be fulfilled;
- the viability of potential solutions and options;
- the implications and impact of each potential solution to the business;
- potential issues for each solution option;
- the risks involved for implementation of each solution option;
- the limitations for each solution option;
- the cost and timescale of developing each option.

When performing a feasibility study, all the inputs and factors that may have an impact on the service are examined. All the operational areas where the service might have an effect will need to be impact assessed. Input into a feasibility study should include (but is not limited to):

- draft service description;
- business and marketing requirements;
- current network topology;
- current IT/business support/service management and OSSs;
- technology experts in the area where the technology may be used for service;
- potential suppliers of any new technology to be used and the potential commercial arrangements;
- forecast of potential end users/customers/usage or network traffic profile.

The feasibility study should assess the impact of the following areas:

- Current network and existing technologies (including network development and enhancement required, capacity planning and other network implications). This should include all network solution options and any new technologies/technology options that might need to be introduced into the network.

- All IT/business support/service management and OSSs implications. Implications on all the system areas as listed Section 8.1 should be investigated. Different system potential solutions or new system technology options should also be investigated.
- All the operational areas that may have process implications. Impact on all the areas listed in Chapter 9 should be explored.
- Interdependencies on other existing or proposed new services. This is especially important if the new service is dependent on additional features to be made available from another service, as this will affect the availability and timescale of the service launch.
- Regulatory implications.
- Interconnections to other operators.
- Potential fraud opportunities.
- Network and system security implications.
- Training requirements for each design option and the cost of training.

In the feasibility phase, service requirements are also clarified and more detailed service require-ments will immerge. These clarifications should be fed back into the product management area for the next version of the service description.

From the above, the output of the feasibility study document should state the following:

- service features and requirements;
- traffic/volume forecast over the next 3 years;
- proposed network design options (stating the network development/enhancements required, pros and cons and limitations of each option and the service feature each option is going to support);
- proposed systems solutions (stating the pros and cons and limitations of each option);
- IT/business support/service management (e.g. billing, the customer relationship management (CRM)) and OSSs (network management, network provisioning, etc.) implications;
- impact on all operational areas – impact should include process changes and potential head-count implications
- capacity planning and network build implications;
- sales support impact;
- fraud implications;
- security implications;
- regulatory implications;
- interconnect and regulatory dependencies;
- interservice dependencies;
- limitations of each options;
- risks, assumptions and potential issues;
- cost profile (CAPEX and OPEX year-on-year breakdown and write-off costs) for each design option. Cost should also include any training cost for sales and operational personnel – cost of potential test platform and environment requirements should also be included;
- estimated timescales for design and implementation of the different options;
- recommendation/conclusion.

5.3 Design and Develop

5.3.1 Gathering Service Requirements

At the start of the design and develop phase, the first and probably most important task is to clarify and define, in detail, the service requirements. As discussed in Chapter 3, gathering requirements can be an onerous task and may not be obvious to many that it is important.

Most service requirements are derived from the service description that is produced by the product manager. The service requirements provide a more detailed description as to what is required to support all the service features for the service. These requirements should contain enough detail for the network, systems and operational teams to define further details for their respective areas. For example, for a broadband service, the service description may say 'the end users can report faults and receive updates on fault progression on the service using the telephone, 24 hours a day, 7 days a week'. To fulfill this requirement, one of the operational requirements may be a team of 30 people, on shift, to support the potential volume of calls for fault reporting. The number of people required is dependent on the estimated fault rates, number of end users and the waiting time for each call before being answered. The means by which the end users can receive updates will also need to be defined. As illustrated above, one requirement from the service description may result in multiple service requirements.

Therefore, when collecting service requirements, it is worth asking the following questions at a high level:

- What are the applications for the service?
- What added value does the service provide apart from providing network connectivities?
- Are there any content provisioning requirements in the service? That is, is providing content part of the service?

  o If so, where will it be hosted?

- What are the service features for the service? This could include:

  o network features (e.g. 2 Mbits downstream bandwidth to handheld devices, or specific network protocols to be supported);
  o operational support features (e.g. 24 × 7 technical support); or
  o system features (e.g. provide fault status updates to end users every 8 h).

- What kind of service quality are the end users/customers expecting (e.g. best efforts or carrier grade or specific delay tolerance levels)?
- Is end user/customer authentication and authorization part of the service?

  o If so, what are the encryption requirements (if any) and message integrity/confidentiality protection requirements?

- Will the service be holding end users/customer personal profiles or data (e.g. identity, personal details, address/site locations)?

  o If so, what are the data protection requirements? (For example: Can the data be passed onto a third party? What security requirements are there to protect this data?)

- What are the service interfaces (i.e. ordering the service, reporting a fault in the service, billing and payment for the service, service enquiries and SLA reporting) to the customers/end users? Is it via the phone, Internet, or in person?

In Chapter 3, the answers to the questions for business, marketing, reporting, security, technical and functional requirements should provide a good basis for service require-ments. Further investigation and questioning of these answers will form more detailed service requirements.

Service requirements are usually broken down into (but are not limited to) the following categories:

- Network requirements:

  o end-user feature requirements;
  o customer feature requirements;
  o network provisioning requirements from the network point of view;
  o network management requirements on the network point of view;
  o network security requirements;

- System requirements:

  o order management requirements;
  o service management requirements;
  o network provisioning requirements from the systems point of view;
  o billing and accounting requirements;
  o bill reconciliation requirements;
  o capacity management requirements;
  o fault management requirements;
  o network management requirements from the systems point of view;
  o traffic and network performance management requirements;
  o service performance, service management and internal reporting requirements;
  o customer reporting requirements;
  o customers' interface requirements;
  o end users' interface requirements;
  o system user interface requirements;
  o external system interface requirements;
  o system nonfunctional requirements;
  o system security requirements.

Other service requirement areas may include regulatory requirements and pricing and cost requirements.

It is suggested that the service designer should work through each of the different categories and break down all the necessary requirements into testable statements. As part of this exercise, interviews and workshops with other internal personnel (e.g. network designer, system designers of existing internal systems) will be necessary to ensure that the requirements are complete and that they are well understood.

Most of the systems requirements are related to operational processes. Both operational processes and the support system requirements go hand in hand. Therefore, when gathering system requirements, the operational personnel must be consulted. Quite often, this can become a 'chicken and egg' situation, where the systems team will say that they need the operational processes defined so that they can capture the systems requirements and vice versa. The operational personnel should specify their requirements on the system at the requirements stage. During the design stage, when the particular system has been designed and the compliance of these requirements will be known. Some of the noncompliances to system requirements will result in manual tasks which form part of the operational process/detail work instructions within the operational teams.

All the requirements should be logged (ideally in a requirements catalogue/database) and traceable with originator. More complex requirements may also contain notes stating why they are necessary or why they have been modified. Once the requirements have been agreed, then priorities will be assigned to each requirement from business and service points of view. Priorities are assigned, as it is virtually impossible to implement all the requirements in one go and not all requirements are needed to the same degree. The most commonly used priorities are:

- 'priority 1 – must have (can't do without)';
- 'priority 2 – required (needed for the service)';
- 'priority 3 – nice to have (nice feature for the customers)'.

This could be a painful task, as, very often, people will tell you that all requirements are priority 1s!

5.3.2 Gathering Operational and System User Requirements

Operational requirements can form part of and will also derive from the service requirements. As illustrated in Section 5.1, some operational requirements are resource related (e.g. additional head counts, desks required) and some are process related. Process-related requirements could be enhancing the existing operational processes or developing new ones for the service. Using the example in Section 5.1, a new team of 30 people may be required to answer fault-related calls from end users. A set of new fault management processes may be needed. The requirements of these operational processes are heavily dependent on the service experience one wants the customers/end users to have when reporting the faults. Therefore, it is important that these operational requirements are captured at the start of the development process in order that the operational processes are defined in a coherent manner with the OSSs to deliver a desirable customer/end-user experience.

Operational processes requirements can be split into the following areas:

- customers/end users sales engagement and support;
- customers/end users order management;
- customers/end users network provisioning;
- customer/end users service provisioning;
- customers/end users service change requests;
- customers/end users service termination and cancellation;
- customers/end users fault reporting and management;
- customers/end users service management;
- network management and maintenance;
- customer/end users billing enquiries;
- billing exceptions;
- bill reconciliation;
- capacity management;
- customer and internal reporting support;
- systems support and maintenance.

Detail operational requirements may result from holding process workshops for all the process areas stated above and interviewing the relevant operational personnel.

5.3.2.1 Systems User Requirements

Systems user requirements (sometimes known as system usability requirements) are not often included as part of the service requirements, but are included as part of the system requirements, or sometimes not included at all in any requirements documents. As part of the operational requirements, system users (normally the operational personnel) will have requirements on the support systems in order that they can deliver the customers'/end-users' experience required for the service. These requirements are not easy to define, as usability is very subjective. However, the risk of not defining system user requirements is great, as the system may not be usable/operable by system users, even though all the system requirements have been fulfilled. For example, the fault management system takes a system user 30 min to navigate through the menu to log one fault due to its complicated menu and slowness of the system. This is virtually not usable from the system users' point of view, and the end users/customers will certainly not have a positive experience when logging a fault. System user requirements are especially important for systems user acceptance testing. This should form part of the operational service testing (as describe in detail in Section 11.1.6), where the users can test the usability of the system before accepting it. This confirms that the system functions are in line with the operational processes that have been developed.

5.3.3 Service Design

After all the requirements have been defined in sufficient detail, the design activities can take place. The main constituent parts of a service are:

- the network that the traffic of the service is carried over;
- the systems solution that supports the service;
- the operational processes that ensure the service is running smoothly.

As part of the service design activity, the service designer needs to ensure all three parts of the solution for the service to work together. Therefore, design workshops of various themes need to be held to ensure that all the design parties are synchronized. Taking the requirements in Section 5.3.1, one might organize a design workshop for fault management. The parties involve will be the fault management systems design team and the fault management operational team. During the workshop, it will be useful to walk through various possible scenarios (or used cases, for those who are more systems orientated) to define what needs to be done within these scenarios and how the systems need to behave to support the system or operational requirements. Further detailed system requirements may result from these design workshops.

After a series of workshops, a high-level design of the service solution should take shape. This high-level service solution should be documented in the technical service solution document. The high-level solution design should detail network components to be used, which systems are used for what system functions and the high-level operational processes to be put in place. The technical service solution document should give a full end-to-end view of the solution and how the solution components will work together as a whole.

Topics to be covered in the technical service solution document include (but are not limited to) the following.

- Network architecture, covering the following network areas:
  - access network;
  - network termination devices (CPE);
  - core network;
  - signaling network;
  - transmission network;
  - other licensed network operator (OLO) interfaces;
  - network management network;
  - service management systems and network.

- System solution and architecture, with system description covering the following system functional areas:

  - customer creation and management;
  - order management;
  - network provisioning and termination;
  - service provisioning, call/session control and service termination;
  - billing, rating and charging;
  - service accounting, revenue reporting, OLO bill reconciliation and revenue assurance;
  - network management;
  - fault management;
  - network and system performance management;
  - capacity and traffic management;
  - reporting – customer and internal reporting;
  - system support and management.

- High-level operational processes:

  - sales engagement;
  - customer services;
  - service and network provisioning;
  - service management;
  - network management and maintenance;
  - system support and maintenance;
  - network capacity and traffic management and network planning;
  - revenue assurance.

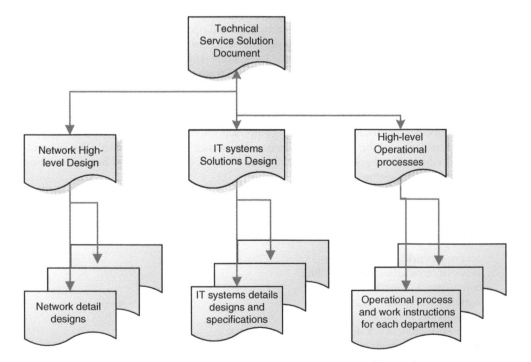

Figure 5.1   Design document structure.

The areas listed above are the constituent parts of the service, known as the *service building blocks* in this book. These will be further explained in Chapter 6. The design considerations for network, system functions and operational processes are covered in detail in Chapters 7–9 respectively.

This high-level solution design should drive the development of the rest of the solution. This high-level solution should not include details of network configuration or detailed design of the systems. This document should refer to other design documents for details on the network, systems and operational process. The design document hierarchy is as shown in Figure 5.1. This high-level solution document should be agreed and signed off by the relevant design authorities before proceeding to the detail design stage.

5.3.4 Network Design

From the network requirements, the service design workshops and looking into the various technology feasibilities for the service, a high-level design/service network architecture should result. The high-level design/service network architecture should describe:

- the network topology for the service;
- the network technologies/equipment used in each part of the network (access, core, transmission network, etc. as described in the service building blocks in Chapter 6);
- physical locations (site/network node) of network equipment;
- the number of network elements per network node;
- protocol to be used for carrying the customers'/end users' traffic across the network, signaling and control, equipment management, session management, service policy management and network service end points/handover points;
- capacity planning rules;
- traffic engineering rules;
- scalability and limitations of the design.

The service network architecture should provide an end-to-end view of the service network and how each of the network elements works together to meet the service and network requirements. A compliance statement of the service and network requirements should also be included in the appendix of the service network architecture document.

Detail network design and configuration will follow the high-level network design, specifying what needs to be built and the configuration of each piece of network element. Internet protocol (IP) addresses, port numbers, etc. will be stated in the detail design document. Service network design conceptual model, network requirements and network design considerations are covered in more detail in Chapter 7.

5.3.5 System Design

Different system analysis and design techniques (e.g. SSADM, UML) can be used for the detailed designs of the systems to satisfy the system requirements. However, what is often missing is the systems architecture/systems solution design for the whole service. The main difference between designing a system solution for a service and a system is that the system solution consists of many systems together which perform many functions to support the service requirements, whereas a system is only part of the system solution. Each of these functions within the service requirements may/can be satisfied by more than one system. Therefore, one needs to decide which system should perform which function.

For example, do you put the order management functions in the CRM systems or do you have a separate order management system? To produce the service management reports, do you build a separate reporting system for all the service management reports or do you build the reporting functions into each service-management-related system? Where should the master data for customers/end users be stored? These are some of the many architectural questions needing to be answered before proceeding to the system design stage for the individual systems.

The systems high-level solution design/architecture should address the architectural questions/issues and dictate which systems will be performing which functions for the service. Without the architectural view of the various systems, there is no means to assess whether the systems provided form a complete system solution for the service. The system architecture should also state the scalability, bottlenecks and limitations of the systems solution. For example, the order management system may be able to process 3000 orders a day; however, the service provisioning can only cope with 2000. Therefore, the maximum number of service orders that can be completed end to end is 2000 per day. A compliance statement to the system requirements should also be included as part of the systems high-level/architecture design document.

This system architecture is normally done by the system architect within the operator/service provider or the system design authority for the service, in conjunction with the service designer. Often, the service design workshop mentioned in Section 5.3.3 can be used to help the formation of the systems architecture. System functions to be performed to support a service are detailed in Chapter 8.

Once the system architecture is formed, the next stage of the design process is to decide what enhancements are required for existing systems and which new systems are required. Requirements specifications are produced for all the systems (new and existing) requiring development. System design is a topic in its own right, and I have no intention to go through it here. Suggested further reading includes *System Design Methodology: From Principles to Architectural Styles* [21] and *System Analysis, Design, and Development: Concepts, Principles, and Practices* [30].

As part of the implementation strategy, one might decide to purchase an existing system off the shelf and tailor it to your needs. Regardless of the implementation strategy, requirements document/system specification for each system is required.

5.3.6 Process Design

High-level process design normally involves running a series of process workshops with various operational support areas. These workshops are normally divided into the different operational work streams (e.g. sales engagement, order management, fault management). In the workshops, the participants will run through the different operational scenarios and the processes to support these scenarios will be documented. This will also result in specifying requirements for detail work instructions.

Before these workshops are to take place, it will be useful to prepare and send out a list of possible scenarios such that the participants of the workshops will have a chance to think them through. These scenarios may not be dissimilar to those of the 'used cases' that were developed as part of the system design activities. It is a good idea, if not essential, to have the relevant system design team present at the workshop as high-level system functions can be within the processes being defined. The high-level processes developed should satisfy the processes requirements defined from the service requirements. Further details on operational processes can be found in Chapter 9.

These process workshops are normally facilitated by the process designer/process design authority for the service. However, the owners of these operational processes should reside within the operational team of each operational area. Detail instructions are written by the relevant operational areas.

5.4 Implementation and Test

5.4.1 Implementation

Once all the designs have been agreed, we enter into the implementation phase where everything within the service is built. At the start of the implementation phase, the implementation strategy should be agreed. This dictates the direction of the implementation activities. Details on implementation strategy can be found in Chapter 10.

5.4.1.1 Network Implementation

From the network perspective, all the network equipment has to be purchased, additional capacity may need to be built and the network implemented as designed. A project plan for building the network will be developed when all the delivery dates of various network equipment are known. Requests for rack space, power and the correct cabling for the network equipment should not be forgotten, as these little items can ruin your project plan!

If new technology is to be introduced to the network, then extensive equipment testing may need to be carried out to ensure that the introduction of the new network equipment will not have a detrimental effect on the current network operation. Please refer to Appendix for the new network technology introduction process. If a new piece of network equipment is to be rolled out to all the network nodes, then, apart from carrying out testing within the new technology introduction process, a network roll out plan will be required. This plan should be the combined effort of the network build department, network operation and marketing department. It may be desirable, for technical and operational reasons, to start rolling out the new network equipment in a certain geographic area; however, from the marketing perspective, those areas may not have many target customers/end users. Therefore, the three parties need to agree on a roll-out strategy.

The network build and roll-out strategy should be in line with the technical and service integration plan to ensure that all appropriate verification, integration and testing have taken place before a large-scale roll out. This is to minimize the risk of damaging the existing network and ensuring that network operations are ready to manage the new network equipment.

5.4.1.2 Systems Implementation

From the systems point of view, implementation will involve writing the system software and code/unit testing the software that has been written. Software tools may be required to implement the required design. Purchasing of system servers will also be part of the implementation. Like network implementation, one should never forget about rack space, cabling, power and air-conditioning requirements to the new system servers. If enhancements are to be made to existing systems, then additional system hardware (e.g. memory, disc space, processing power) may be required.

Owing to timescale (or other logistical reasons), the system software may be delivered in different phases for different functionalities. The challenges here are the integration of the various systems with different software versions and the configuration management of the systems.

When all the software and hardware deliveries of the various systems are known, a project plan for the systems implementation should emerge. That should be fed into the technical and service integration plan.

5.4.1.3 Operational Implementation

Operational implementation will involve:

- finding and/or recruiting the required operational support personnel for the service;
- finding desk spaces for new support personnel (if required);
- writing detail work instructions in accordance with the processes that have been defined;
- training the support personnel on the new service and the processes they need to use to operate and support the service.

If a new or big operational team is required (e.g. a new customer service team), then one may also consider outsourcing the function to a third party. This is, of course, dependent on the cost effectiveness and quality of the service requirements.

5.4.2 Integration and Testing

Integration and testing is split into five main parts:

- unit/detail testing of code and/or configuration;
- technical testing of individual network components and systems;
- technical integration and testing of network and systems;
- technical service integration and testing of end-to-end service;
- operational service testing for the end-to-end service.

The technical testing of individual network components and systems ensures that the network equipment and the systems individually perform the required functions as specified in the network and system requirements.

Technical integration and testing confirm that the different network elements work together to deliver the service network features required and that end-to-end network connectivity for the service can be achieved. Similarly, for the systems side, the technical system integration ensures that all the systems work together to deliver the systems solution required. The last stage of technical integration and testing is the service integration of the systems and network together to form the end-to-end technical service solution as specified in the technical service solution document and service requirements.

The operational service testing verifies that the service solution is working well as specified by the service requirements and that all the systems, network and operational processes deliver the desired customer/end-user service experience.

Different stages of integration and testing and the details of service integration strategy are covered in Chapter 11.

5.5 Service Launch and Operate

One of the main ingredients for launching a service successfully is the operational readiness for the service. Therefore, as part of the service integration and launch phase, the operational service testing is the key activity to ensure that all the operational processes defined have satisfied the operational requirements, the systems user requirements are met and that the operational personnel are trained.

Operational service testing involves running through various possible operational scenarios to ensure that all the operational processes are in place and that the training given to the support personnel is sufficient. This also confirms that the whole service works end to end. For example, one operational scenario to test could be an end user placing an order for the new service. This order should go through all the network, systems and operational processes from order management to network provisioning, service provision and billing. Further details on service integration and launch can be found in Chapter 11.

When the service is in its infancy, teething problems will arise. Therefore, additional attention may be required and the design team should be made available to support the resolution of any operational issues. Post-service launch reviews will be a good forum to facilitate the resolution of any outstanding issues.

5.6 Service Withdrawal

Service withdrawal is not a topic many people think about, mainly because it is not very exciting and often not a priority within the company. Most people would rather spend their time thinking about building something new and wonderful. However, like launching a new service, service withdrawal needs to be managed, thought through/designed and implemented to avoid any potential risk and liabilities there might be.

Withdrawing the service could be for various reasons:

- Technology replacements, where new technology supersedes the old one and new services are launched to replace the old service.
  - In this scenario, customer/end user migration will need to be thought about.
- The operator/service provider decides that the service is no longer profitable and the service should be withdrawn.
  - In this scenario, customers/end users are migrated in another service or terminated from the service.
- The service has come to the end of its useful life.

Topics to think about when withdrawing a service may include (but are not limited to):

- customers'/end users' contractual commitments;
- network equipment maintenance agreements;
- systems maintenance agreements;
- informing customers/end users on the service withdrawal plan;
- customer/end-user migration to the new service (if relevant);
- management of decoupling network elements with the rest of the network;
- management of decoupling systems from the other systems that they are integrated to;
- removal of functionality with the systems where the systems are shared with other services;
- disabling the service from the service management systems (e.g. CRM, billing, fault management and reporting systems);
- removal of external interfaces (e.g. Web interface for customers placing orders for the service, Web interface for receiving service management reports or interfaces to external supplier or to other network providers);

- removal of marketing material (e.g. list on the operators'/service providers' Web sites);
- safe decommissioning and removal of network equipment – this includes the removal of network elements from the network management system;
- safe decommissioning and removal of system equipment – this includes the removal of systems from the system management system;
- disposal/reuse of network and system equipment;
- network and system resource reassignment;
- operational personnel reassignment.

Bearing in mind all of the above, the first document that the service designer should refer to is the technical service solution document. This should give a good overview on what and how the service was built. Steps to decommission various parts of the service can be identified from there. Further details on service withdrawal are discussed in Chapter 12.

6

Service Building Blocks

Designing a service is like designing a house: both require certain fundamental building blocks. When designing a house, certain considerations (e.g. the orientation, space, size, owner requirements) are essential. This is not dissimilar to designing a service, certain design considerations are required. Therefore, when designing a service, it is very important to understand these fundamentals and consider all aspects for the service. This chapter establishes all the fundamental building blocks of all telecommunications services and explains what a service is made of, from both technical and operational points of view. Examples of a basic PSTN service, managed data services and 3G mobile services are used to illustrate how various fixed and 3G mobile services fit into this service building block model. Additional challenges when designing mobile services are also discussed.

6.1 The Building Blocks

The main building blocks for a network-based service, in the broadest terms, are as follows:

- Network infrastructure on which the service is run. This is where the traffic of the service is carried. The network architecture for the service states all the network elements required to make up the service network and describes how each of the network elements is design and configured to support the service.
- Systems by which the service is run or operated. The system solution provides functionalities that are required by the service, as well as carrying out the business and operational support functions for the service. The system solution architecture/design describes which systems are used and how each of the systems fulfills the service, business and operational requirements.
- Operational processes by which the service is operated. The operational processes and detailed work instructions ensure the service is running smoothly. They describe the steps and the procedures which operational personnel should follow when carrying out manual tasks. The high-level operational processes for the service describe how the service is operated. Lower level procedures and detailed work instructions are combined to describe the manual tasks to be carried out and steps and procedures to follow for each task.

Details within each block are explained in the sections that follow.

Successful Service Design for Telecommunications   Sauming Pang
© 2009 John Wiley & Sons, Ltd

6.2 Conceptual Network Architecture for Fixed and 3G Mobile Services

The conceptual network architecture for all network services is shown in Figure 6.1. The constituent parts are the:

- access network;
- network termination devices/CPE;
- core and aggregation network;
- signaling network;
- transmission network;
- other license network operator interfaces.

Other service enabling network components include:

- network management network;
- service management systems network.

When designing a service, all the above networks are to be considered and will require designing. Design considerations for the network are further explained in Chapter 7 (Figure 6.1).

6.2.1 Access Network

The access network is like the door to the house; that is, as the means to get access to the network services. The access network is typically described as the 'last mile' of the network. It is the network that links the end users' terminals/network termination devices/CPE with the last network access

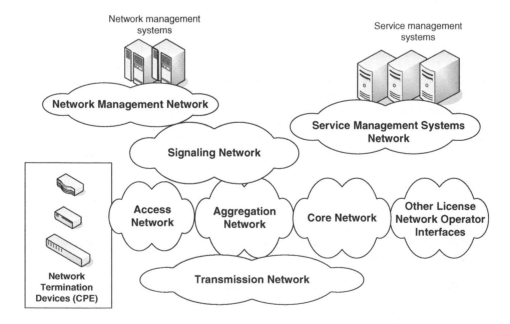

Figure 6.1 Conceptual network architecture.

point of the service provider/operator. Normally, in a fixed network operator environment, it is the network to a local exchange building (the exchange building being the last network point of the operator). In the mobile network environment, it will be the wireless connections to the end users' devices from the base station.

6.2.2 Network Termination (Interface) Devices/CPE

These are devices or equipment where end users/customers connect to the service provider/operator network/facilities. These devices enable users and customers to use the service they have subscribed to. Network termination devices are normally located in the end users' homes or customer premises for fixed-line services; hence, the term CPE is used for these network termination devices. In the world of mobile services, they are mobile handheld devices/handsets. For some services, these are the network demarcation points for the operators/services providers. The network demarcation points are where the service for the operators/service provider ends.

6.2.3 Core and Aggregation Network

The core network performs the switching and forwarding function, such that the data and voice traffic for the service can get to its destination in the most effective way possible. Sometimes, it will be more efficient to aggregate traffic from various access points or networks before passing to the core network. This network is called the aggregation network. It is used to improve efficiency in a large-scale network. In the voice world, a network of tandem switches is a good example of an aggregation network. If an aggregation network exists, then edge switches or routers will link the aggregation network with core network.

6.2.4 Signaling Network

This network carries all the command and call control signals for all calls in the voice/circuit switched network. Signaling System No. 7 (SS7) is the ITU standard specifying how the switches should exchange control commands, alerts, addressing and transmission information across the CS network. Separate and dedicated signaling networks or channels are built for the CS networks, where it carries no customer/end-user traffic. The one exception to this 'rule' is the mobile text messaging service (or otherwise known as the short messaging service (SMS)) where the spare capacity in the signaling network is used to carry customers'/end users' text messages. Signaling in the data/PS network is embedded within the protocol used for the data to be transmitted. For example, the session initiation protocol (SIP) is used for call/session control for VoIP calls/data. No separate signaling network or channel is built for data/PS networks.

6.2.5 Transmission Network

Like the foundations of a house, the transmission network is the base of any telecommunication network services. The transmission network is the optical transport layer of the network that carries the data and voice traffic for the service from one access point of the network to another/its destination. These are sometimes called the 'pipes' or the 'plumbing' of the service. The technologies used for the core network transmission are normally synchronous digital hierarchy (SDH)/ synchronous optical network (SONET) coupled with dense wavelength division multiplexing (DWDM) technologies.

6.2.6 Other License Network Operator Interfaces

Now that the telecoms market is no longer a monopoly, even if you are an incumbent operator, it is unavoidable that you need to connect to the network of OLOs/alternate network providers to exchange network traffic. If you are an alternate network provider or a mobile operator, these interfaces are vital to your survival. Without them, you will not be able to deliver the network traffic for the service to its destination. Traditionally, these interfaces are for voice traffic. In the Internet world, this will be the peering points (e.g. at Tele House and Telecity in London) where you have all the peering arrangements to exchange traffic. There is an increasing demand to exchange data traffic over other data interconnection points.

6.2.7 Network Management Network

How do we know if the network elements are functioning properly? How do you know the network elements are performing as expected? In order to manage the network elements efficiently, the network management systems (normally based in the network operations center) is used to monitor the status of the network. This system needs to be connected to the network element to obtain network status and performance information. The network connecting both parties is the network management network. Quite often, this network already exists for other services that are currently running. However, there may be requirements to have a separate network management network for the service.

The decision to create a new infrastructure for the new service depends on the existing infrastructure the requirements of the new service. Questions to be answered when designing the network management network include: Do you just add the new network elements into the existing infrastructure? At what point do you create a new network? If existing infrastructure is used, how do you ensure the traps from this service are not mixed with the others? How do you enable and maintain a single view of the service and monitor the 'health' of the service effectively? All these will be further explained in Chapter 7 (Section 7.3.4).

6.2.8 Service Management Systems Network

By service management systems, I mean the systems that enable functions for the services and those used to manage and operate the service (i.e. the business support systems as well as service support systems). For example the system that controls the policy and profiling of users, the system that provisions the customers and users on the service network and systems that do reports for customers and so on. Some of these systems can potentially be accessed by customer/users directly (e.g. customer reporting systems or billing systems) and some will require interactions with the network, for example provisioning system. Since the service is reliant on these systems communicating with each other and the systems communicating with the network elements, a network needs to be designed and built to enable this to happen.

6.2.9 Example Illustrations

To illustrate how the conceptual network architecture described above fit within the real life services, below are some example network architectures for different services to place it into context. Network design considerations for all the examples below are further detailed in Chapter 7 (Section 7.3).

6.2.9.1 Broadband Service

Figure 6.2 is a broadband network architecture example. In a typical broadband service, the end-user devices are digital subscriber loop (DSL) modems/router (plus a micro-filter). The access

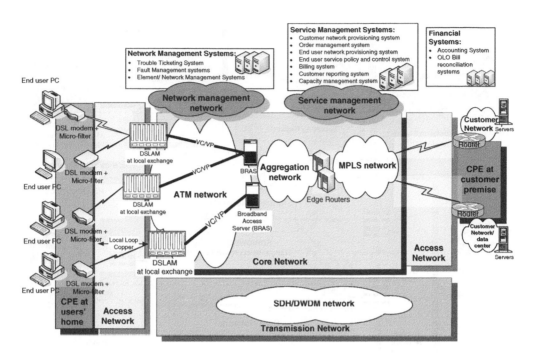

Figure 6.2 Example broadband service network architecture.

network consists of the copper local loop and the network equipment at the edge of the access network comprising of the digital subscriber loop add–drop multiplex (DSLAM) at the local exchange and the core network is either asynchronous transfer mode (ATM) or IP network. Most first-generation DSLAMs have an ATM interface onto the core/aggregation network; hence, an ATM network forms part of the core/aggregation network. However, the new generation of DSLAMs (or generally known as multiple services access switches) are going to have IP interfaces, so the core/aggregation network will be an IP instead of an ATM network. The aggregation network brings traffic from various access points around the country to the core IP network. The technology used for the IP core network can be multi-protocol label switching (MPLS). The core network is used as a delivery network for service traffic to customer/ISP sites. The underlying transport for the aggregation and core network is the SDH/SONET transmission network. DWDM technology is used to make the transmission network more efficient.

To manage all the network equipment, a network management network is built to link all the management ports of all the network elements for the service to the network management systems. This enables all the network events and traps to be viewed by the network operation personnel and for them to resolve any network fault remotely where possible.

For customer/end-user service provisioning and management purposes, a service management network is built connecting the network elements/element managers and the service management/end-user policy management systems. This enables all the interactions between the end-user enablement and the network element to take place. Since all the service management systems need to communicate with each other, this service management network provides the facility to do so (Figure 6.2).

6.2.9.2 PSTN and VoIP Services

Figure 6.3 illustrates an implementation of the network architecture for the PSTN and VoIP service.

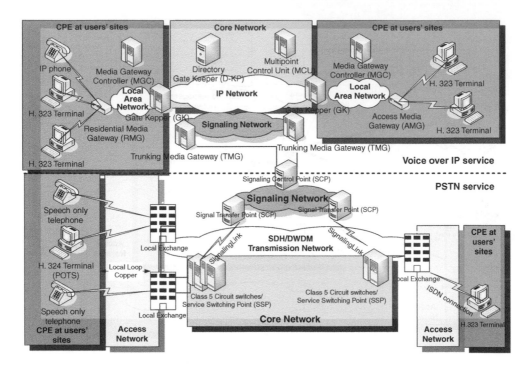

Figure 6.3   Example VoIP and PSTN service network architectures.

6.2.9.2.1 PSTN Service

With the PSTN service, the CPE will be the telephone or the private branch exchange (PBX). The access network work will be the copper local loop between the exchange buildings from the main distribution frame (MDF) to the phone point in the users' houses or local PBX in a commercial premises. Signaling for the voice calls (SS7 signals) is separated at the circuit switch, in the exchange building, acting as the service switching points (SSPs). The signaling is carried over the signaling network, normally separate trunks within the network, to the service control point (SCP) where the routing for the call is decided. The signal for the call is carried over the signaling network all the way through to the destination SSP/circuit switch. The voice part of the call will be carried over the switches, specified by the signaling, across the core network and the access network to reach its destination. Depending on the size of the network, an aggregation network of some tandem switches may be used to collect and distribute voice calls from various geographic areas.

Details of the network architecture for PSTN services can be found in *Guide to Telecommunications Technologies* [6].

6.2.9.2.2 VoIP Service

For VoIP service, the CPE can be IP phones or H.323 compatible terminals. The access network can either be the local area network (LAN) infrastructure for voice to be delivered within the customers' VPN or a broadband access network. The media gateways (MGs) in the access network convert analog voice into IP packets and act as an interface between the CS and PS networks. It also handles traffic prioritization and enforces traffic (QoS) profiles/policies. The interface between the core and access network are the gatekeepers (GKs). The GKs perform look-ups for phone numbers and names to IP addresses and also zone bandwidth management function. The GKs ensure that there is enough bandwidth across the wide area network (WAN) for the call. If the WAN lacks bandwidth to support the call, then the GK can deny the connection request, and ensure the calls currently in progress are not compromised. Directory gatekeepers (D-GKs) are the dial plan database for all GKs. The multipoint control units (MCUs) are used for conference calling and mixing multiple audio and video streams together. The signaling protocols for VoIP can either be H.323 or SIP. The trunk media gateways interface with the SCP in the SS7 world for voice 'breakout' calls where calls need to be delivered to the PSTN network. The media gateway controller (MGC) enables all the media gateways to exchange and translate call signals and control information with each other so that voice and video packets are properly routed through the network. The MGC also controls the information dissemination to, and manages all, media gateways. The core network will deliver the voice data packets in accordance with the QoS policies specified. As always, the transport 'pipes' for these voice packets are the transmission network (Figure 6.3).

6.2.9.2.3 3G Mobile Service

For the 3G mobile service, the conceptual network architecture, as illustrated in Figure 6.4, is no different from that which has been described in Sections 6.2.1–6.2.8. The network termination devices are the UE (e.g. mobile handsets, PDAs, laptops). Within the UE are the universal SIM

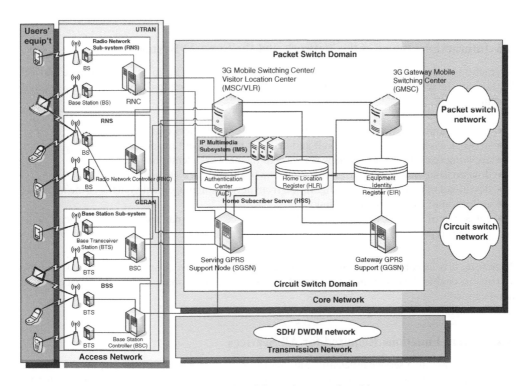

Figure 6.4   Example 3G mobile service network architecture.

(USIM; SIM = subscriber identity module) application, which contains the logic required to unambiguously and securely identify the user, and the mobile equipment (ME), namely the mobile handset itself. The access network acts as the wireless interface between the UE and the core network. The access network in the Universal Mobile Telecommunication System (UMTS) allows two types of access system: the base station subsystem (BSS) and the radio network subsystem (RNS). The core network can be connected to one or both systems. The access network helps to provide the flexibility to keep the core network technology fixed, while allowing different access techniques to be used (e.g. code division multiple access (CDMA) or wideband CDMA (WCDMA)). Within the core network, the following functions are performed:

- mobility management, where the tracking the location of the UE is performed;
- call control, where the establishment and release of voice calls are controlled;
- switching, where the voice connections are established and calls are being delivered to the end points;
- session management, where the data transfer between the UE and end points is established and released;
- routing, where the data packets are routed from the UE to the end points;
- authentication, where users using the service are authenticated;
- equipment identification, ensuring that the handset for the services is genuine.

The transmission network provides the transport from the UE to end points for both voice connections and data transfers. Further details on network architecture for 3G mobile can be found in Section 7.3.3 (Figure 6.4).

6.3 Interactions Between the Network and the Support Systems

Interactions between the network and the support systems form an integral part of the service. Without it, the service will not exist. In the broadest terms, interactions between the network and systems include:

- network and service activation and deactivation for customers/end users;
- network events and status for fault and network management purposes;
- applying fixes, upgrades and new software to the network elements;
- downloading new service features/parameters to the network elements;
- download of new/amended configuration from network management systems to the network;
- collection of call records for billing;
- collection of network data for network performance, traffic management and capacity management reporting.

Therefore, it is important not to forget to design the network that connects the network and the systems to enable the communication stated above. Without a robust design, the service will be handicapped.

6.4 System Functions Required for All Services

In order to operate a service efficiently, automation plays an important role. Hence, systems are built to enable service automation and to minimize human errors. Systems are built to support operational processes and these operational processes are reliant upon functions to be performed by the systems in order to manage and operate the service. These functions should be automated as much as possible to achieve maximum efficiency.

The building blocks for system functions include:

- customer creation and management;
- order management;
- network provisioning and termination;
- service provisioning and call/session and service termination;
- billing, charging and rating;
- service accounting, revenue accounting and OLO bill reconciliation;
- network and service management;
- fault management;
- performance management;
- capacity and traffic management and network planning;
- system support and management;
- reporting.

The specific functions and actions to be performed by the systems are, of course, dependent on the service requirements. However, the above functional areas should be considered when designing systems to support a service. The subsections that follow provide brief explanations on each of the building blocks listed above and some of the functions to be performed within each building block. Further details on functions to be performed within each building block are discussed in Chapter 8.

6.4.1 Customer Creation and Management

To enable the sales department to manage the existing and potential customers, the systems need to provide the facilities to identify customers. Therefore, customer information, like name, address, contact person and their role in the organization, target services for the customers, potential revenue, sector of business and so on need to be held. This will help the sales team to follow up on potential customer leads, as well as to keep track off existing customers and the services (and ideally revenue generated) currently enabled. In addition, customers are to be created before anything can be sold to them. Therefore, the systems need to enable operational personnel to create a customer and enter and modify details of a customer. The above functionalities may also be applied to end users. However, the scale of the operation will be much higher.

Customers may also need to be created in other customer-facing systems; for example, customer reporting systems and billing systems if orders are being placed. End users' accounts may also require creation, depending on the service being offered.

6.4.2 Order Management

Order management is primarily concerned with how the orders are being progressed once the customers/end users have signed up for the service. This involves monitoring and tracking of order status and their progression through the network and service provisioning cycle/process. Order status is updated at various points of the provisioning cycle. Therefore, meaningful status should be defined per service at various trigger points of the network and service provisioning cycle. To monitor the progress and track failures, timeframe and jeopardy flags are to be set on each order or order progress trigger points for SLA purposes. These failed orders will require human attention.

Orders can be of different types; for example, new service installation, customers/end users moving to different location, additional sites or end users to the same accounts. All these need to be defined as part of the service description and are separated at an early stage of the order management process, as they may require different treatment and management.

As part of the order progression, the availability of network and system resources is to be checked and reserved to ensure successful network and service provisioning. Depending on the service being provisioned, the system may need to be able to identify the network capacity and system resource requirements and reserve the necessary capacity for order fulfillment.

6.4.3 Network Provisioning and Termination

Network provisioning functions ensure the customers and the end users have end-to-end network connectivity for the service to be activated. This enables them to access and receive information/data required. System functions in the network provisioning and termination/deprovisioning areas can be broken down into the following categories:

- define network connectivity to be provisioned or terminated/deprovisioned for each customer/end user order;
- check network capacity availability for each customer/end-user order;
- flag potential/actual capacity shortfall;
- provision/activate customers' end-to-end network (deactivate network connections for network terminations);
- provision/activate end users' end-to-end network (deactivate network connections for network terminations);
- ensure/test end-to-end connectivity for each customer/end-user order;
- update network inventory for all the network connections or disconnections made;
- update the status of the orders.

For most telecommunications services, the transport 'pipe' for information/voice/data is the network. Therefore, without the network infrastructure being activated, no service can be provisioned.

6.4.4 Service Provisioning and Call/Session Control and Service Termination

Within the service provisioning, call/session control and service termination areas, the system functions are related to end-to-end service enablement. The functions may include customer service activation with the appropriate service profile (e.g. security management, privacy management, QoS policy), reporting and billing functions and so on. For end users, services may be activated at a different level. For example, different end users may have the following differences:

- credit policy and price plan control;
- security and authentication policy;
- session/call control policy;
- QoS policy;
- service packages that define some of the different policies (combination of all the above) that can be used for end users.

For services with a large number of end users, it is essential to have a system that manages the different end-user profiles. The effective management of activation and/or termination/deprovisioning of network and system ensures efficient use of resources for the service.

As part of the service provisioning (or termination) activities, reporting and billing accounts will require activation (or termination), when the service and network provisioning (or termination/deprovisioning) activities are completed. For a positive customer service experience, the customers should be informed of the service activation and the date when billing for the service will commence. At service activation, reporting data is to be collected and compiled. For service termination, the customers should be sent the final bill after successful service termination.

6.4.5 Billing, Rating and Charging

Billing and charging of customers/end users seems to be conveniently forgotten by most. Quite often, it is not dealt with until the last minute. However, the billing and the charging functions are vital for the cash flow of the company. The basic system functions in this area should include:

- defining the charging structure for the service;
- creation or termination of customers/end users' accounts;
- collecting billing information (e.g. call detail records (CDRs), service activation and so on);
- rating of calls or sessions (if required);
- creating invoices per customer/end user;
- dispatching invoices;
- collection of payment;
- support billing enquiries and disputes.

Other additional functions may be required, depending on the billing requirements for the service. For example, applying discount schemes, invoice to be sent out on a different date per customer request, choice of quarterly or monthly bills and so on. More details will be covered in Chapter 8.

6.4.6 Service Accounting, Revenue Reporting, OLO Bill Reconciliation and Revenue Assurance

Product managers normally have a revenue target to meet for their service portfolio. Therefore, service accounting and revenue reporting will be required. Service accounting should account for the revenue on a per service basis, whilst revenue reporting may be from a per customer perspective.

Inter-operator accounting and reconciliation (the term OLO bill reconciliation is to be used in this book) is related to accounting for and reconciling of payment between operators. This is to ensure records for the calls/data sessions between operators are accurate and that they are rated correctly. This involves comparing the bills from other network operators (OLOs) with the internal records. This is to prevent unnecessary outgoing payment to other network operators. If there are discrepancies between what has been charged and what has been used, then exceptions should be raised for human attention.

Revenue assurance functions are associated with reports and activities to ensure accuracy of revenue to be collected and the revenue that is due.

6.4.7 Network and Service Management

Network management functions are concerned with managing the events and alarms generated by the network, as well as managing the performance of the network. With effective management of network events, faults in the network should be fixed or prevented before customers/end users raise fault tickets. Network management system functions include:

- monitoring the status of the network;
- collecting network events;
- correlating network events into network alarms;
- raising network alarms in the system to the appropriate priority levels;
- tracking planned outages;
- providing the ability to reroute network traffic due to congestion or around major failures.

Operational processes will need to be defined to resolve network faults as it is not realistic to build a network management system that can automatically diagnose and fix network faults. Hence, many requirements for network managements will come from operational processes for the network management area.

Service management system functions are mainly concerned with service performance (and sometimes may include network performance) reporting, managing customer SLAs and monitoring internal service KPIs. The actual management of the service is reliant on the service management operational processes and the actions to be carried out from the results of the service management reports. Basic service management system functions may include:

- assessing service impact as a result of network faults;
- analyzing service performance from network performance data (e.g. data error rate, network throughput data from network interfaces);
- monitoring customer SLA and issuing jeopardy warning where customer SLAs are about to be breached;
- defining, producing and delivering internal and external (customer) service (and network performance) reports.

6.4.8 Fault Management

Fault management in the context of this book is managing faults raised by the customers/end users and faults raised internally (e.g. faults discovered by the network management systems). A fault management system is a tool to help with tracking fault tickets (also known as trouble tickets) raised both internally and externally. The fault management system can potentially be a customer-facing tool. It can also be a workflow management tool for managing customer-raised faults. This is dependent on the service requirements and the organization of the operational environment. The main functions for a fault management system include:

- logging fault tickets;
- tracking progress of fault tickets;
- updating fault tickets status;
- logging fault resolution/diagnostics and resolution timescale;
- alerting operations of SLAs that are about to be breached;
- SLA reporting on fault resolution and exceptions.

6.4.9 Performance Management

System functions for performance management are mainly divided into three areas:

- network performance;
- system performance;
- application performance.

6.4.9.1 Network Performance

Network performance management of the network is essential to achieve good service performance and customer/end-user perception of the service quality, as well as to optimize and maximize the network asset of the company. The system functions in this area are mainly:

- defining performance data to be measured;
- defining report format and threshold levels;
- data gathering;
- network traffic report generation and monitoring of threshold levels;
- traffic profiling;
- network modeling and simulation.

With the predefined algorithm and data collected from the network, reports are generated so that the network planners are able to spot the traffic 'hot spots' in the network and, in turn, plan and build necessary additional network capacity and resources accordingly. Therefore, most of the requirements for traffic and network performance management systems will come from network planners and capacity management functions. These requirements may not be service specific, but based on network node locations or on a specific network element. However, it is not unusual to have service or customer requirements on certain network performance data/reports, especially when QoS parameters are involved, with associated SLAs and service level guarantees (SLGs) on the QoS parameters.

6.4.9.2 System and Application Performance

Similarly, for system performance management, the system functions are the same. However, the performance measures are related to systems performances; for example, CPU utilization, memory usage and so on. These performance measures are used by system managers to ensure that the systems are healthy and are performing as expected.

The application performance management system functions deal with the collecting of application usage date, monitoring of application transactions and measuring predefined application performance parameters. These measurements ensure the application is performing as expected.

6.4.10 Capacity and Traffic Management and Network Planning

The functions for the capacity management tool are similar to that of the network performance management system. Reporting and simulation from the network model/tool should identify potential capacity shortfall/surplus in the network given the current usage levels and traffic forecast figures. The traffic management system uses the network performance data to route traffic where capacity is available to optimize the network utilization, hence optimizing network cost. Support system capacity functions should also be included, as the support systems are very much part of the service.

Depending on the service requirements, potential or actual capacity shortfall may also be reported (e.g. port shortage on a certain piece of network equipment) during the network provisioning cycle. Capacity shortfall may or may not be acceptable depending on the service's SLA for service provisioning, cost and the lead time of network and system capacity. Therefore, requirements for the capacity management function are going to vary between services.

6.4.11 Reporting: Customer and Internal Reporting

Reports can be generated for customers, mainly for SLA purposes, and for internal users to monitor the health of the service/service performance. The basic functions for reporting systems are:

- defining data to be collected and interval of data collection;
- defining reporting format and medium of reports;
- collecting data;
- providing temporary storage of data;
- performing required (defined) manipulation or analysis;
- presenting results in graphical or tabular format;
- storing and archiving reports as required;
- distributing reports to audience.

This makes it sounds so easy. In fact, reporting is probably one of the most difficult items to develop. Reporting requirements are not always apparent or considered at the start of the service design activities. Although, customer reporting requirements are apparent as part of the service description, this does not always represent a full set of reporting requirements for the service. For example, KPI reporting for the service, traffic management reports, capacity management reports and so on may not be identified until later. Many of these reports are not captured until it is too late.

There is also a perception that, if the data is there, reports can be generated. This is not entirely true. Reporting requirements may have huge implications on the database structure of the system to be developed. In addition, there is also the mechanism of delivery of the reports to be considered. Additional network or security infrastructure may result. Therefore, it is essential to capture as many valid reporting requirements as possible at the start of the service design process to avoid huge and potentially costly system changes. In Chapter 8, example reports are listed in each of the functional areas for consideration.

6.4.12 System Support and Management

System support and management functions are concerned with managing the events and alarms generated by the systems. This is to ensure that all the systems are functioning in an efficient manner and that the system availability for the service is kept. Like network availability for the service, system availability for the service is just as important, as this can form part of the customers' SLAs and affect the customer/end user experience of the service. Hence, this is an area that cannot be neglected.

With effective management of systems events, system resources and regular maintenance, faults in the systems should be spotted, fixed or prevented before customers or system users raise fault tickets. The main functional areas for the systems support and management systems are:

- monitoring the status of the systems;
- collecting systems events;
- raising alarms in the monitoring system to the appropriate priority levels;
- tracking planned outages;
- measuring utilization of system components (e.g. disc space, CPU);
- measuring system performance (e.g. CPU response times under a specific load).

Systems events are normally dealt with by IT operational teams. Operational processes will need defining to resolve system faults. Like network management requirements, many requirements for system support and management will come from operational processes for the systems.

6.5 Operational Support Processes for All Services

One might ask why we need operational processes. If we have the network and the system to operate and support the service, then doesn't everything just happen automatically? In an ideal world, the system will perform all the tasks required to operate a service and the network will not fail, but in reality this is hardly the case. Operational support processes are required for both the systems and human actions.

Operational support processes define the sequence of events and tasks to be performed to fulfill certain operational/service requirements. These processes are normally end-to-end processes where there are defined inputs or events and defined outputs with a sequence of tasks to be performed to complete the required operations/operational functions. For example, the operational support process for end-users' order management will have end-user orders as input and completed orders as output with the order management operational support process that defines the tasks and the sequence of events that are required to fulfill the end-user orders.

Operational support processes are made up of two elements:

- High-level processes that define what needs to be done to perform certain operations. These define the sequence of actions and events to fulfill the service requirements at a high level from start to finish (end-to-end process).
- Low-level procedures and detailed work instructions that define, in detail, the procedure and steps required to be performed by operational personnel to complete certain operational functions within the high-level processes

These processes are defined in conjunction with system functions. The OSSs are designed to perform operational tasks to achieve efficiency; therefore, the operational support processes should be defined in parallel with the system functions. Not all tasks can be performed by the systems, either due to system constraints or human decisions being required. Therefore, when designing a service, you need to define the tasks to be done and the processes (i.e. sequence of events) by which these tasks are performed before deciding which tasks are to be performed by the system and which tasks are to be carried out manually. For services that are using existing systems, the tasks and processes are to be defined with the functionality of the existing systems in mind. Where appropriate, systems enhancements may be required. Detailed work instructions will need to be defined for all the manual tasks.

Detailed work instructions define the actions and steps required to perform certain operational tasks to be carried out by people or human tasks to be performed under certain operational scenarios. These step-by-step instructions/guides ensure all the right actions are carried out in the correct sequence. These are normally written by the respective operational departments after the system designs have been finalized.

One also needs to distinguish between operational support processes and business support processes. Operational support processes are a sequence of tasks that will operate and support the service (e.g. order management process, fault management process). Business support processes are processes that will enable the business/company to function (e.g. recruitment process, supplier management processes) irrespective of services. The processes covered in this section and in Chapter 9 are limited to operational support processes for the service. For business support processes, please refer to the eTOM [31].

The processes described below (and in Chapter 9) are within the operations area in the eTOM, but with more details and emphasis on service operations. The exact mappings to the eTOM will be discussed further in Chapter 9. The categorization of these areas is based on operational functions rather than operational departments, as each organization is different. Therefore, when designing the service, the organization of these processes might change, but the processes themselves should still be relevant. As always, these processes should be adapted to the specific service you are designing.

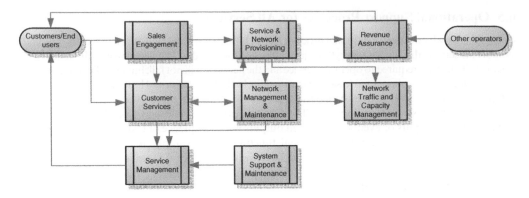

Figure 6.5   Interrelationships between operational processes.

There are eight main operational support process areas for a service (Figure 6.5):

- sales engagement;
- customer service;
- service and network provisioning;
- service management;
- network management and maintenance;
- network traffic and capacity management and network planning;
- system support and maintenance;
- revenue assurance.

Figure 6.5 shows the interrelationships between the different operational support processes. Details describing the interrelationship can be found in the sections that follow.

6.5.1 Sales Engagement

Sales engagement processes mainly focus on helping the sales team to sell the service. The main item to be produced for the service will be the marketing and sales literature. This should include all the features, functionality and limitations of the service. Depending on the service, a press release and website description may also be required.

The sales engagement processes should include items like:

- customer relationship management;
- customer leads tracking and customer information management;
- up-selling to existing customers/end users;
- customer qualification and credit checking;
- pre-sales and post-sales customer support.

As part of the pre-sales support, the customer network and service profile design and customer network order feasibility should be performed before contract signature. This is to ensure that the service can be provided and that there is network capacity available at the required locations.

6.5.1.1 Inter-Process Relationships

Sales orders should result as part of the sales engagement process. This is fed into the network and service provisioning process. Information from the customers/end users is fed into the customer service process for future reference.

6.5.2 Customer Service

Customer services processes and work instructions are designed to ensure that the customer-facing operational teams have all the information and handling procedures available for:

- customer/end-user service enquiries – ensuring all the answers are available to the customers/end users;
- customer/end-user complaints – ensuring all the customer complaints are dealt with appropriately;
- logging faults raised by customers/end users and updating customers with fault resolution actions;
- customer/end-user billing enquiries and disputes;
- customer/end-user service and network termination;
- customer/end-user service change requests and so on.

As part of the customer services processes and work instructions, it may be a good idea to define ways to retain customers/end users on the service if customers/end users want to cancel or terminate the service. Understanding why they want to leave the service is also worth capturing as part of the work instructions. The opportunity for up-selling to customers may also be defined as part of the customer service processes, as well as in the sales engagement process.

6.5.2.1 Inter-Process Relationships

Most customer service operational support processes are initiated by the customers/end users. For customer/end-user change requests and service/network termination requests, the processes that are followed will be in the service and network provisioning area. For faults raised by the customers, the fault management processes should feed into the network management and maintenance processes for fault diagnostics and resolution. Depending on the service and how the operational teams are organized, fault resolution and update may be provided as part of the service management process.

6.5.3 Service and Network Provisioning

Service and network provisioning processes deal with tasks and events from the point when the orders have been accepted to the completion of service and network provisioning of customer/end-user orders. Part of the order acceptance may include order validation, if it is not done as part of the sales engagement processes. The service and network provisioning processes should include processes that deal with:

- customers/end-users network order management;
- customers/end-users service order management;
- customer network implementation (including customer service testing);
- customers/end-users network provisioning;
- customer/end-users service provisioning.

6.5.3.1 Inter-Process Relationships

Service and network provisioning processes have interfaces into:

- revenue assurance processes for customer/end-user billing following the successful completion of service and network provisioning activities for the customers'/end-users' orders;
- network management and maintenance processes from network inventory update and network configuration point of view following successful completion of service and network provisioning activities;
- network traffic and capacity management processes for network capacity if network capacity shortfalls have been discovered as part of customer/end-user network provisioning processes.

6.5.4 Service Management

Service management processes are concerned with how the service is managed effectively. It should ensure that the customers/end users are happy with the service. Within the service management processes, there should be processes to monitor service performances; for example, monitoring current open faults raised by the customers, monitoring service KPIs and customer SLAs to ensure the service is performing as expected and so on. Processes around customer SLA reporting and performance reporting should also be included in this area, because these are the indicators on service performance and how the customers will measure the level of service received. If service degradation has been detected as part of the service monitoring processes, then a feedback process for the various operational departments to improve the performance should also be established.

As part of the service management processes, processes should be included to ensure the customers/end users are informed of any service affecting planned engineering works and service outages, to improve customer/end-user experience and avoid logging unnecessary fault tickets.

There will be times when the customers/end users are not happy with their service and want refunds or service credits. Criteria for application of pricing adjustments, refunds, discounts, rebates and issuing service credits should also be part of the service management process.

6.5.5 Network Management and Maintenance

Network management and maintenance processes define how the network should be monitored, managed and maintained. The processes should include:

- *what* needs to be done when network faults occur (e.g. opening fault tickets, problem diagnostics and troubleshooting, corrective actions, service restoration and closing fault reports);
- *who* should be informed;
- *which* support agency should be resolving the network faults and so on.

As part of the processes, network disaster recovery processes should also be included.

For proactive network management, network maintenance procedures, network performance management processes and network configuration management work instructions should also be defined.

In order to ensure that the network faults are rectified at the earliest opportunity, effective management of fault resolution resources (e.g. internal resources in the field to fix the faults) or third-party support agencies are required. Therefore, essential are processes/a framework to monitor the performance of third-party support agencies and fault-resolving resources in the field.

6.5.5.1 Inter-Process Relationships

The main process interfaces for network management and maintenance are:

- service management processes for informing the customers/end users on service affecting network faults;
- customer service processes to inform customers/end users of known faults when customers/ end users try to log faults.

Other process interfaces include network capacity and traffic management as a result of monitoring network performance or rerouting of network traffic due to fault conditions.

6.5.6 *Network Capacity and Traffic Management and Network Planning*

Capacity management processes should define how the network capacity should be managed after reviewing all the network performance/traffic and network utilization reports. The capacity management processes are only meaningful if working in conjunction with the network planning and network building processes. Whatever the network capacity requirements are, they need to be planned and built. That is when network planning and building processes take place. The planning rules for the network equipment/elements are defined as part of the network design for the service and will be of use for capacity management and network planning purposes.

6.5.6.1 Inter-Process Relationships

Traffic monitoring and management processes are input into the capacity management processes. Traffic management processes should ensure the network traffic is routed to where network capacity is available without incurring too much inefficiency in the network. Traffic monitoring, network capacity management and network planning processes form an iterative cycle to ensure the network assets are maximized without compromising the service performance.

6.5.7 *System Support and Maintenance*

System support and maintenance processes define how the systems are supported and maintained. The processes should include: system monitoring and system fault management – handling, tracking and resolution of system faults. Detailed work instructions for troubleshooting, fault diagnostics and performing corrective actions for each of the systems should be stated in the system support and maintenance detailed work instructions. System disaster recovery procedures should also be defined as part of the system support processes, with detailed work procedures and instructions to cover the recovery of all the systems.

For proactive system management, the system maintenance, system performance monitoring and management and procedure (including monitoring the performance of third-party support agencies) should also be defined. As for proactive system management, the system configuration management of the systems is also very important to avoid unnecessary system faults/service outages. System security procedures should be part of the company's security policy rather than being defined on a per service basis, unless the service has specific security requirements.

6.5.7.1 Inter-Process Relationships

The main interface for system support and maintenance is the service management processes relating to system performance for the service and system faults/outage/planned works that are affecting the service. The system support and maintenance processes have secondary interfaces to all the other operational support processes where system users relay on the support systems to perform their operational roles. Processes should be set up to inform system users of any planned works or system outages.

6.5.8 Revenue Assurance

Revenue assurances processes ensure all the revenues due for the service are collected and any unnecessary outlay for the service is captured. Billing fulfillment and customers'/end-users' payment collection for the service are very much part of the revenue assurance process. The process should include the collection of customer/end-user debt and overdue payments.

Potential outlays for the service could be renting the access network from OLOs for fixed-line services or end-user roaming charges for mobile services. In both cases, reconciling the bills presented by the other network providers with internal call records is important to avoid unnecessary out-payments.

6.5.8.1 Inter-Process Relationships

Interfaces to the revenue assurance processes are mainly the service and network provisioning processes for revenue collection, bill reconciliation with OLOs and the customers/end users for revenue and debt collections. OLO bills are the input into the bill reconciliation process.

6.6 Are there Any Differences Between Fixed-Line and 3G Mobile Services From a Service Design Point of View?

Fundamentally, the concepts and the building blocks of the service (as illustrated in the sections above) for both fixed-line and 3G mobile services are much the same. The major difference is in the access technologies (as stated in Section 6.2.9.3 and later in Chapter 7) and the mobility capabilities it brings. The technological differences and the mobility capability do present different challenges for the service designers. Below are some of the differences and the associated challenges:

- **More complex re-sale channels**. Mobile services are primarily sold via reselling channels rather than directly sold by the service providers. The billing and commissioning of sales can be more complicated than the direct sales fixed-line services.
- **Emphasis on personalization**. The direction for mobile services is personalization of the services with self-service capabilities (i.e. having the capabilities to change the personal and service profiles via portal interfaces). Provisioning of portal services can only be effective with centralized service control architecture and with a fully integrated back-end OSS, in relation to order processing, service provisioning and so on.
- **Cultural challenges in IT operational environment**. With the centralized service control architecture there are cultural challenges in the IT operational environment. Traditionally, IT/OSS functions are only in the 'back-end' support domain and are not involved in the call/session set-up (i.e. 'live' traffic environment). With centralized service

control architecture, the service profiling tools and session control systems at the heart of the call/session set-up and processing. The traditional IT support function is now critical in handling 'live' traffic. Hence there is a cultural shift from being the 'back-end' support to 'front-end' operations.

- **Access network challenges**. The dynamic allocation of resources at the air interface leads to a different set of design and planning rules for the access network. The interoperability testing of mobile devices with RAN (similar to CPE testing against access network equipment for a fixed network) will become more important as a greater variety of applications are to be enabled on mobile devices. The service providers will need to ensure that the service/application can be enabled on the end-users' mobile devices with the correct operational software. Additional challenges also include the verification of service settings and a mechanism of upgrading the operational software on mobile devices to enable new services.
- **Service creation tools**. Many of the application interfaces within the UMTS framework (e.g. open service access (OSA)) have been standardized as part of the 3rd Generation Partnership Project (3GPP) [9]. This makes the network more accessible to the service/application providers and third parties. Services can also be created with various toolkits that are available. This simplifies the service creation/service design activities. Although the concepts and system functions for designing services detailed in this book still apply, the service development cycle and time to market for new services can be reduced.
- **More complex value chain**. With the network services more accessible, the value chain for mobile network providers, services providers and application/content providers is now more closely integrated. This introduces complications with billing and revenue apportionment with the various parties in the value chain, especially for those services with application transactions outside the service provider environment (for example, end users downloading content from third-party providers) where the value chain has been extended to the third party. This can be further complicated by commission and end users making these transactions while roaming to another mobile network. Therefore, with all the above complexity, the billing functions will need to be content/application aware. With a more integrated and complex value chain, there is also a higher risk of fraud and revenue leakage.
- **Content management**. The trend for all service providers, especially for mobile service providers, is to own the content to make their service offerings more attractive. To make this profitable and efficient, an effective content storage and management tool is required.

6.7 Summary

Figure 6.6 is the summary of all the service building blocks for a service. Essentially, telecommunication services are made up of the:

- *service network* carrying the network traffic for the service;
- *systems* with their functions to support the service and its operation;
- *people* who operate the service with defined *operational support processes* to ensure all the necessary procedures and tasks are performed correctly.

These building blocks are independent of technologies and can be applied to all telecommunication services, regardless of whether they are fixed or mobile services. Some adaptation and tailoring may be required to fit specific service requirements. However, the basic concepts and principles apply.

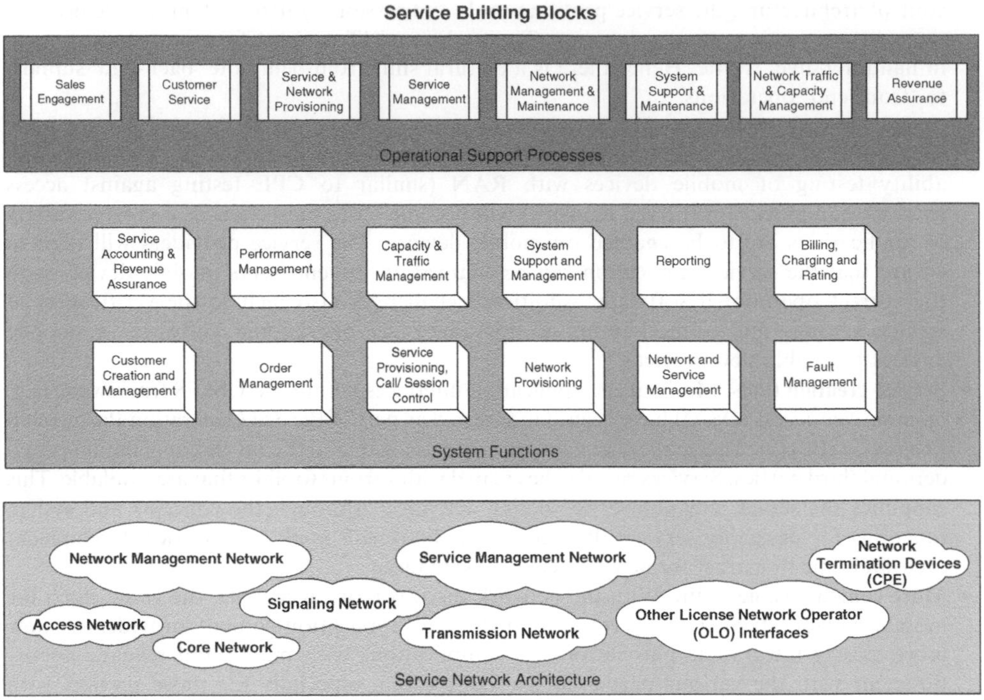

Service Building Blocks

Figure 6.6   Summary of all service building blocks.

6.7.1 Service Network

The service network can be broken down into the following components:

- access network;
- core and aggregation network;
- transmission network;
- signaling network;
- customer/end-user network termination devices;
- OLO interfaces;
- network management network;
- service management network.

All the network-based services will have at least some, if not all, of the above components. The only components that might not be relevant will be:

- the signaling network for PS services, as the signaling for data packets is part of the protocol used and no dedicated signaling network is required;
- the interface to OLOs, as not all services (e.g. IP VPN service) require such interfaces.

For application-based service, all the above network components may be transparent/hidden, as the Internet may be the network used. However, the Internet does have these components, and if any part of the component is not in working order, then your customers/end users will suffer from service degradation. Service network architecture is further discussed in Chapter 7.

6.7.2 Support Systems

As for the systems that support the service, the following functional areas should be considered when designing the service:

- customer creation and management;
- order management;
- network provisioning and deprovisioning/termination;
- service provisioning and call/session and service termination;
- billing, rating and charging;
- service accounting, revenue reporting, OLO bill reconciliation and revenue assurance;
- network and service management;
- fault management;
- performance management;
- capacity and traffic management and network planning;
- system support and management;
- reporting – customer and internal reporting.

The system functions required to support the service are very much dependent on the service requirements and the scale of operation. When designing a service, it is worth going through each of the functional areas to ensure that all the scenarios and functions are considered. However, if not all the functional areas are relevant for the particular service you are designing, then it is still worth thinking it through to ensure none of the system functions has been missed. Some of these functions may be carried out by operations personnel rather than by the system. Further detailed system functions are covered in Chapter 8.

The main roles of the system are to make the operation of the service efficient and to keep operational cost low. Therefore, when developing the systems, it will be a good idea to bear in mind the cost and benefit of the systems, as well as the potential cost of human error.

6.7.3 Operational Support Processes

On the operational support processes front, the following operational areas for a service should be considered:

- sales engagement;
- customer services;
- service and network provisioning;
- service management;
- network management and maintenance;
- network traffic and capacity management and network planning;
- system support and maintenance;
- revenue assurance.

The operational process groupings/areas are dependent on the organizational structure of the company you are designing the service in. In some cases, for example, you may merge the customer service and service management process areas together. Or you may also separate the fault management process into a separate process area, rather than grouping it under the customer service area. You should tailor these processes as required. However, the areas/processes listed should be part of the service, whether it is known or grouped under a different name or operational area.

There is not always a direct mapping between system functional areas and operational processes. As explained in this chapter, both of them are closely linked and should be defined in parallel, as the purpose of both the system functions and operational support processes is to operate and support the service. The system functions are developed to improve service efficiency and reduce operational cost.

The conceptual network architecture, system functions and operational support processes described in this chapter form the basic building blocks for the service. These building blocks and design principles are discussed further in Chapters 7–9.

If one stretches the mind a bit further, the service building blocks stated here are relevant to all services provided by any service provider of any service sector. Each building block just needs to be put into the context of the service being provided. This is especially true for the transport and logistic or postal sectors, where these sectors provide services to transport people or goods from one place to another, whilst the telecommunication services enable the transport of information from one place to another. Drawing the parallel with the transport/logistic sector, the service network in a telecommunication service is equivalent to the transport network. In a railway/train service, the network is the railway network; the network technology is the technology of the trains being operated; the signaling network for these trains is much the same as that in a telecommunication service; the customer services processes and network management system functions are also very similar. With this, I shall leave you to ponder which other industries these service building blocks can be applied to (Figure 6.6).

7

Network Design and Development

New services often result from innovations in new technologies or applications. The network and mobile terminal technologies and designs that support the services are crucial parts of the service. Quite often, new technologies are introduced to the network as a result of new services and new business requirements. Apart from the network that carries the services requiring design and development, there are other supporting networks which are also required but often forgotten until too late.

In this chapter, only general network design concepts and considerations are discussed, as network design is heavily dependent on the technology involved. Various fixed and 3G mobile services are used to illustrate the service network building blocks (as described in Chapter 6) and any additional design considerations for each of the services/technologies involved for each of the service network building blocks. Network requirements, network security, service configuration on network elements, network planning and capacity planning are also discussed in this chapter.

The new network technology process explained in Appendix is designed to present readers with a methodical approach to introducing new technology into the network with minimal risk. This is useful for new services that require new network technologies, and this process is designed to run in parallel with the new service design process described in Chapter 4.

7.1 Network Requirements

Before any network designer can start designing, network requirements need to be known. Often, the network requirements are derived from service requirements. Other sources of requirements include current network constraints, physical network element locations, existing OLO interconnect constraints, cost constraints and operational requirements and so on. A good way of capturing requirements is asking questions. Below are some of questions that need answering before any network design activity should take place. This list is by no means exhaustive.

7.1.1 Service Network Requirements

- What are the current services and applications to be supported (e.g. e-mail, video, voice, file transfer, database updates and transfer)?
- What are the anticipated future services and applications to be supported?
- What are the service features that need to be supported?
- Are there any user security/authentication requirements?
- What is the source and destination of traffic?
- What are the line-speed requirements?
- What line-speed variations need to be supported?
- What are the traffic profile, direction and routing and so on?
- What networking protocol does the service need to support (e.g. file transfer protocol (FTP), simple mail transfer protocol (SMTP), real-time transport protocol (RTP)? Are there any other specific routing protocols that the network needs to support?
- What is the network interface presentation (physical and protocol layer) to the end users and customers?
- What are the resiliency requirements for the customer site delivery network?
- Does the network termination equipment/CPE need to be managed?
- Is contention or compression acceptable?
- What signaling protocols need to be supported?

7.1.2 Network Volumetric and Capacity Planning Requirements and Network Constraints

- What are the volume and characteristics of the traffic flow?
- What are the network utilization constraints?
- What is the forecasted volume of traffic and over what timescale?
- What is the network utilization that will make the business case for the service work?
- What is the geographic coverage of the service? What type of coverage is required?
- What geographic area will the calls/data packets need to be delivered to (national, international or just local)?
- For mobile services, what is the spectrum available for use?
- What are the cost constraints on the network capital spend?
- What are the constraints (location and physical line speeds) for interconnections with OLOs?

7.1.3 Network Performance Requirements

- Are there any traffic shaping, queuing or QoS requirements?
- What is the acceptable time delay/latency between the source and destination of the calls/data packets?
- Are there any fragmentation requirements for the data packets?
- Will the customers/end users pay for the network performance requirements specified?
- What are the network reliability requirements?
- What are the network performance requirements? What are the drivers for these performance requirements?
- What KPI does the network need to support? What data does the network need to provide for KPI reporting?
- Are there any network SLAs that need to be supported for the service?

7.1.4 Network and Service Management Requirements

- What are the network management functions required? What network monitoring and management protocol are to be supported?
- What network management protocol does the network need to support (e.g. simple network management protocol (SNMP))?
- Does the service require remote access to the network elements? If so, over what mechanism, and what security measures are required?
- What is the network management architecture?
- What are the service management functions required?
- What are the network security requirements?
- What are the disaster recovery requirements?
- What are the network availability requirements?
- What data does the network need to provide for billing, SLA management, network performance reporting, capacity planning and management and customer reporting?

The level of detail of requirements depends on the level of detail design that needs to be produced. Capturing requirements and producing designs are normally an iterative process. Most design activities start with some high-level requirements. As the design activity goes on, more details need to be defined. In order to maintain control of the design process, it is recommended that the designers should base their design on a set of controlled and signed-off high-level network requirements. Any additional requirements should be under change control. Otherwise, the design activity will take forever and achieve nothing. Please see Section 3.1.4 for requirement definitions and attributes.

7.2 Technical Network Considerations

With the network requirements above clarified, further technical network design considerations can be considered.

7.2.1 Network Topology

The network topology defines the physical and logical network configuration. There are four basic types of network topologies:

- Star. This is where the network links are connected to a central node. This is commonly used in a client—server environment, where most of the data need to be communicated back to the central server.
- Mesh. The network elements are interconnected directly to each other. This topology offers resiliency between network elements; however, scalability could be an issue when the size of the network expands beyond a certain limit.
- Ring. All network elements are connected in a logical and/or physical ring. This topology is commonly used for the SDH transmission network. The token ring LAN environment also uses this topology.
- Tree and branch. This is where many network connections are branched out from a single network node. This is commonly used in the access network or cable networks where there is a single path between all elements.

7.2.2 Network Architecture

The network architecture sets out how the network operates and functions. It defines how various network elements communicate with each other and the protocols used in/compatible with the chosen architecture. Considerations for selecting the network architecture include:

- The type of application using the network. Depending on the requirements for the service, particular network architecture may be selected because it facilitates applications that must be supported by the network.
- Type of device being used in the network. There might be a specific requirement for the use of a specific device. That might dictate the network architecture being used.
- The physical topology. The physical network (cable and ducts) that is already in the ground. It may not always be possible to start from scratch when design and building a network.
- Traffic forecast requirements. The volume of forecasted traffic may influence the network architecture. To fulfill the future demand and growth, the network architecture may move towards a certain direction.
- Performance requirements. The performance requirements may influence the choice of network architecture and network technologies to be used.
- Resiliency requirements. To support the resiliency and network availability requirements, the network may need to be architected in a certain way to fulfill these requirements.
- Technological direction. The network architecture may be dictated by the type of technology/ standard used. For example, for UMTS services, the network architecture is pretty much defined as part of 3GPP standard [9].

Some suggested further reading on network architecture is the book *Network Analysis, Architecture and Design* by James D. McCabe [2].

7.2.3 Network Element Selection and Node Location

Selection of a specific device/technology depends on many factors, which may include:

- the ability to fulfill the service/network requirements;
- the network functionality supported by the network element;
- the protocol/signaling that needs to be supported;
- the cost of the network element;
- compatibility with the existing technologies that are already in the current network;
- manageability of the network element;
- support arrangement with certain suppliers, etc.

7.2.3.1 Node Location Selection

The major deciding factor for the physical location of the nodes is the location of sources and destinations of network traffic. Normally, nodes are placed near the major sources and destinations of traffic. However, this might not always be economical or physically possible. For example, there might be a major source of traffic (e.g. Internet service) from site A (e.g. Tele City), but there is no switch site there. Therefore, it is more economical to use the transmission network to bring the traffic back to a switch site location than setting up a new site for the node near site A. More often than not the ideal site for the node equipment may be full and it is physically impossible to put additional equipment in there anymore, so alternative sites need to be found.

The other factor for consideration will be the location of network interconnections with OLOs. You might need to place a node near the network interconnections to obtain or send the traffic to other operators. Balancing out the cost effectiveness between placing an additional network node close to network traffic sources and the additional cost of the transmission network to the current network node locations is the key to node location selection. The physical node location is one of the most difficult, complex and costly aspects of network design and planning. Getting it wrong can be a very costly mistake.

7.2.4 Node Sizing

Sizing of the network is heavily dependent on the traffic forecast estimated by product management and sales functions. Without it, it is rather difficult to size the network nodes and the network in general.

Another key influencing factor is the function of the node – if it is a transit node or access node, or both. Although these functions are logically separate, they are physically important in node sizing.

The other major factor for sizing of the node and the capacity of the network is the traffic flow. This can be estimated when the volume of traffic, the network topology (hence flow of traffic), the traffic profile (peak traffic rate versus normal), and the type and mix of traffic are known.

Other deciding factors for sizing the network include the:

- processing and memory power of the node (quite often, the type of traffic and the traffic mix have a strong influence on this);
- protocol being used;
- amount of simultaneous sessions a node can handle;
- throughput of the node.

The network utilization requirement is also an important factor. Only high network utilization can achieve network cost efficiency. Please see Section 7.2.10 for further traffic engineering and capacity analysis techniques to achieve network efficiency.

7.2.5 Phone Numbering, Domain Name and IP Addressing

Phone numbering and IP addressing are the most common mechanisms that enable the network to identify a user/users' network termination equipment uniquely. Phone numbers are used in the voice/circuit switching world, while IP addresses are used in the data/packet switching world and domain names are used in the Internet world.

7.2.5.1 Phone Numbering

Phone numbers/E.164 names are defined by the ITU-T recommendation *E164: The International Public Telecommunication Numbering Plan*. In general, there are two broad numbering schemes: geographical numbering and nongeographical numbering. Geographical numbering has number ranges for different geographic areas. For example, 0207 will be the number range for central London while 0208 will be for Greater London. This enables efficient routing and data filling in the switches. For nongeographical numbering, different services are separated by different number ranges, for example 07xx numbers are generally for mobile services, 0800/1-800 are free phone/toll-free numbers and 09xx numbers are for premium-rate calls. Different number ranges have different charging rates for the calls.

All number ranges are subdivided for different service providers/OLOs to use. So each operator may also have a number range in a certain geographic area. These number ranges are data filled into the voice switches in accordance with the network configuration such that the calls are routed to their destination most effectively (either geographically or to another operator). Typically, the routing for the voice calls is fixed. There is always a primary and a secondary route for each number range from each voice switch.

7.2.5.2 Domain Name

Another way to identify users can be in the form of 'username@domain' as defined in *RFC 1035: Domain Names – Implementation and Specification* [64]. The registration of the top-level domain (TLD) (e.g. com, org, uk) system is administered by the Internet Corporation for Assigned Names and Numbers (ICANN). They decide if, how and when new TLDs are added. Domain names are registered in domain name services (DNSs) servers and typically maintained by service providers, where the domain names are 'mapped' to the IP addresses of the destinations/the server where the end users are connected.

7.2.5.3 IP Addressing

For IP addresses, on the other hand, the story is slightly more complicated. IP addresses can either be public or private. For the Internet, public IP addresses are to be used. They are subdivided into class A (range 1.xxx.xxx.xxx to 126.xxx.xxx.xxx), class B (range 128.0.xxx.xxx to 191.255.xxx.xxx) and class C (range 192.0.0.xxx to 223.255.255.xxx). These ranges only apply to IP version 4 (IPv4) (RFC 791) [3]. Although for public IP addresses the class A IP addresses are divided into different geographic areas around the world, any class B or C IP address is typically nongeographically related. Therefore, routing the data packet efficiently to its destination requires an effective routing strategy (see Section 7.2.6). Private IP addresses are normally for internal use only. Therefore, it is not uncommon to consider using private IP addressing schemes for internal network equipment, especially as IPv4 addresses are rapidly running out. IP version 6 (IPv6) (RFC 2460) [58] addresses are becoming the new standard, with a 128-bit length address, and are a prerequisite for a UMTS network and services.

Depending on the service being provided, IP addressing of the end-users' network termination devices may be fixed or dynamically allocated. For example, for a broadband service, it may be more secure for the users to have their IP address dynamically allocated every time when they initiate a session. This will reduce the chances of being spoofed or threatened by denial-of-service attacks and is an efficient way of using the scarce IP address resources. IP addresses for the network nodes are normally fixed to make routing updates simpler.

For mobile services, the mobile IP technology/framework is used. The mobile terminals have home (IP) addresses assigned and used in the home network. When the mobile terminals are in foreign networks (i.e. roaming to a different network), care of addresses is assigned to the mobile terminals. Please see Section 7.3.3.4 or refer to *Mobile IP Technology and Applications* [8] or *Mobile IP Technology for M-Business* [33] for further details on mobile IP technologies.

For VoIP services, having the correct IP addressing scheme is crucial, as the number of IP addresses required could be large. For enterprise networks, it is recommended that private IP addresses are used, just due to the sheer number of addresses required (well, until IPv6 comes along anyway). In the voice and data convergence world, inter-working with the PSTN network is inevitable, and the mapping of IP addresses to phone numbers is essential for these converging networks. The telephone number mapping (ENUM) concept (RFC 2916) [14] takes a step further in proposing a domain name system-based (DNS) architecture and protocols for the mapping of phone numbers to a set of attributes (e.g. e-mail address or URL) that can be used to access an Internet service. Please refer to ENUM (RFC 2916) [14] for the latest details.

7.2.6 Routing Strategy

The more efficiently you route the traffic across the network; the higher the network utilization and greater cost efficiency can be achieved; hence, you can provide more competitive services with the same network asset. There are three different routing schemes you can use:

- one route or multiple routes between nodes;
- minimum hop/least cost routing versus minimum distance/shortest path routing;
- fixed/static routing versus dynamic routing.

It is very unusual to have only one route between network nodes, as this will not fulfill the network resiliency requirements. The one-route strategy will probably be used when there is only one neighboring node being connected. Multiple routes between nodes are more common.

The minimum hop routing scheme will send the traffic through the least number of immediate nodes, while minimum distance routing will send the traffic over the shortest possible path.

Fixed/static routing sends the traffic down only one route for a particular node, irrespective of the network conditions, while dynamic routing may change the route of the traffic depending on the network condition (i.e. if there is a link failure, then the routing of the node will be changed and traffic will be sent to a different route, or if a certain route is congested, then the node will send the traffic through another path). Fixed/static routing is the simplest form of routing.

Dynamic routing enables the network to route the traffic according to the latest network condition. The network status information is exchanged amongst network devices. For IP networks, there are two methodologies for updating the state of the network: distance vector (minimum hop/least cost) and link state (shortest path). The routing information protocol (RIP), interior gateway routing protocol (IGRP) and open shortest path first (OSPF) are the commonly used dynamic routing protocols. RIP and IGRP are dynamic routing protocols for distance vector routing, while OSPF is a dynamic routing protocol for link state routing. The convergence time (time required to update and reconfigure the routing table) of different dynamic routing protocols also varies. Please refer to *Designing and Developing Scalable IP Networks* [12] for details on routing strategies.

For voice services, the most common routing strategy is least-cost routing, and the SS7 keeps track of the link states of the voice network. Hence, the SSPs can dynamically reroute traffic to avoid congestion or network link failures.

The actual routing scheme to be used depends on the application and the technologies being deployed. Different parts of the network within a service may use different routing strategies to make the network more efficient.

7.2.6.1 QoS Considerations

Unlike the circuit switching world, where time slots/network capacity are dedicated and guaranteed to each end-users'/customers' session, the packet switching environment (namely data networks in general) allows the network resources to be shared without dedicated/guarantee capacity for each end-user/customer session to achieve higher efficiency. However, the downside to this is that all data traffic types are treated equally. In order to offer differential services (i.e. delivering delay-sensitive traffic, e.g. voice and video, before other data traffic that have higher delay tolerance, web download) within this environment, different quality levels within the packet switching environment are to be defined. The different quality levels (known as QoS) define the traffic treatment mechanism set that attempts to guarantee consistent levels of service performance and yet maximize network utilization and achieve efficiency.

QoS requirements are dependent on the services being designed. QoS is typically defined by the following parameters/characteristics:

- bandwidth/data rate/capacity;
- transfer delay;
- delay variation;
- bit error;
- packet loss.

Different router scheduling and queuing mechanisms are used to deliver data traffic within the guaranteed limits within each classified QoS level to achieve different quality levels The following are some common scheduling and queuing mechanisms:

- **First in, first out queuing**. The data packet arriving first in the queue will be first out in the queue. This mechanism suits data traffic requiring only best-effort delivery. It is not suitable for applications with high QoS requirements (such as video or voice traffic), as the router will start discarding data packets once the queue is full regardless of traffic class or type.
- **Priority queuing**. Data traffic is prioritized (into high, medium and low) by filtering on source and destination addresses or application port numbers. High-priority data packets will get preferential treatment; however, under congestion conditions, medium- and low-priority data packets can wait a long time for delivery.
- **Class-based queuing and weighted round robin**. Similar to the priority queuing mechanism described above, data packets are prioritized and put into their respective queues. Each queue is dealt with equally in rotation in a round-robin fashion. The size of the queue (in terms of byte size) determines the rate at which the data packets are accepted in the queue. This does not work well with mixed traffic types, which have packet sizes that vary between large ranges.
- **Weighted fair queuing**. Similar to the priority queuing mechanism described above, data packets are prioritized and put into their respective queues. A different weighting is assigned to each queue, where high-priority queues are served more frequently than lower priority ones. This type of scheduling is a resource-hungry operation and deemed more suitable for low-speed edge operation.

For full details for the different scheduling and queuing mechanism, please see *Mobile IP Technology and Applications* [8] and *Designing Networks with Cisco* [57] for details.

Different signaling and control mechanisms are also required to accomplish the desired end-to-end QoS. ATM technology has a very comprehensive definition of traffic class and QoS delivery mechanism. However, ATM signaling is complex and network equipment is expensive. Hence, most favor IP-based technologies. Below are very brief descriptions of some common and Internet Engineering Task Force (IETF)-recognized IP-based protocols/delivery control mechanisms to achieve QoS:

- **Integrated Services (IntServ)**. Intserv (RFC 1633) [38] is based on network resource reservation upon the request of the application. It accepts or denies the request base on the network load at the time of request. IntServ provides two levels of service:

  o guaranteed service (RFC 2212) [39] – has upper limits for queuing delay, but no limit on delay variation;
  o controlled load (RFC 2211) [40] – a better than best-effort service.

Resource reservation protocol (RSVP) is the signaling protocol that enables user applications to request a specific QoS from the network. More details on IntServ can be found in Refs [12, 38–40].

- **Differentiated Services (DiffServ).** DiffServ (RFC 2475) [41] provides the mechanism for classification and conditioning of data traffic. Similar traffic types (this could be same type of traffic, same source and destination address, same direction of flow or same port number, etc.) are grouped/classified into different streams. The classified traffic is metered and measured according to its agreed traffic profile. Appropriate markings (DiffServ Code Points (DSCPs)) are applied for treatment at the next hop. Shaping is applied to ensure the different streams conform to the predefined profile and that bursts of traffic are 'smoothed out' by delaying the packets.

  The forwarding treatment of packets with certain code points is called per-hop-behavior (PHB). There are three different service levels provided by PHB:

  - Expedited forwarding (EF). EF (RFC 2598) [42] tries to guarantee the delivery of data packets. This can be used for low delay, low jitter and low error rate applications.
  - Assured forwarding PHB. Assured forwarding PHB (RFC 2597) [43] has a less stringent guarantee on delay and error rate and allows different levels of service to be offered. These are defined in the RFC.
  - Best effort. No priorities are provided for the data packets.

  More detail on DiffServ can be found in Refs [12, 41–43].

- **Multi-Protocol Label Switching (MPLS)** (RFC 3031) [16] protocols enable network traffic to be forwarded or switched using a simple label (as detailed in RFC 3031). This is a forwarding protocol mainly applied at the edge of the network and not an end-to-end service guarantor. It can run over IP, ATM, frame relay, Ethernet and IP networks. The labeling technology added the connection labels to the connectionless IP network. Instead of analyzing the address and routing to the next hop, the MPLS technology adds a label (according to the destination IP address, host address and host address QoS) to the data packet and forwards it to the next hop. Different service class labels are applied within the MPLS labels and are used in the next hop to specify the ongoing hops. Through the use of DiffServ and IntServ techniques (described above) and the ability to perform traffic shaping and engineering, different QoS and class of service (CoS) can be achieved. More detail on MPLS can be found in Designing Networks with Cisco [57].

A decision as to which mechanism and protocol to use for the service depends on the QoS requirements for the service or application being designed and the complexity of the network involved. In reality, it is likely that combinations of the above mechanisms are used to achieve the desired QoS requirements. Cost implications to implement such QoS requirements should also be a consideration.

7.2.7 Resiliency, Disaster Recovery and Business Continuity

There is always a requirement to build a resilient network. What does that mean? And where does it come from? The network resiliency requirement normally comes from the network availability requirements for the service. Having a resilient network helps to achieve the required network availability figures. From the *Oxford Advanced Learner's Dictionary* [1], resilient means 'able to feel better quickly after something unpleasant'. However, trying to work out a network availability figure from the MTBF figures from each of the network elements supporting the service is very difficult. A good network design should have no single point failure.

One way of designing a resilient network is to have the capability to automatically reroute traffic or self-correct in the event of network link or network node failures. Therefore, it is a good design principle to have alternative routes available for the traffic to go without a service outage under a network failure scenario. When sizing the network nodes, this is an important factor to consider. Ideally, the network should be designed in such a way that, if any of the network nodes has gone down, the traffic will be dynamically rerouted to the neighboring nodes and there should be enough capacity in the neighboring nodes to absorb the traffic. Of course, this may not be possible due to cost constraints. However, it is a design principle one should adopt where possible.

There may also be a requirement to operate a service across different geographic areas. For example, the service might be operated across multiple data centers. In this case you might decide to facilitate the switch-over should one data center fail. Alternatively, you can load balance across the sites or dynamically select a site depending on traffic level. The data center systems synchronization/mirroring strategy has a huge bearing on the method for rerouting.

Network diversity in the WAN, especially in the transmission network, can also help to achieve network resiliency, especially in the scenario of rerouting the traffic around a different network/transmission path. This will also work well in the access network, where the traffic is terminated at the customer's site. Diverse fiber, for example, can avoid service outage if one of the network links to the customer's site is broken. Sometimes, the back-up line, especially to the customer's site, can be of a different technology. For example, for a 10 Mbit/s leased line to a customer's site the back-up network could be an ISDN/xDSL link. Although the service has been degraded, the cost of having a diverse link is much reduced compared with another 10 Mbit/s leased line.

The number of network nodes also plays an important role in network resiliency. Although having fewer network nodes at fewer geographic locations will be easier to manage and, hence, reduce the operational cost, this could be very high risk when it comes to resiliency. For example, if a city is only served by a single switch and there is a catastrophic event at that switch site, then the service to that city will come to a complete standstill. If that city were served by several network nodes, then the service may be degraded or not affected by the catastrophe at all.

With a resilient network, disaster recovery activities need not be a real-time recovery issue. Well, at least you are not losing any traffic and, hence, potential revenue, as, at a minimum, you can provide a degraded service. The ease of recovery for the network node from a disaster depends on how much of a disaster it is (i.e. how much damage was caused). The key to disaster recovery is to have the network node configuration back up onto the network management system regularly. The network management system should also reside physically away from network nodes. In the scenario where the whole network node is destroyed, once all the hardware and software have been replaced, it should be relatively easy to recover the network node after re-installing the service configuration (one that was on the network node before the disaster). However, without the back-up the process will be a bit of a headache, as you need to reconfigure everything from scratch and the chance of missing something is very high; therefore, this is not recommended.

The other consideration is the business continuity/recovery plan for a major disaster. How would the business continue and maintain the essential services at the crisis location? What are the network design considerations to support the business continuity/recovery plan? Depending on the business objectives for business continuity, there may be requirements to have network nodes mirroring each other or some network data (not just service configuration on the network node) need to be backed up or synchronized. In addition, there may be requirements for the network management network to support multiple network management system locations in the event where the primary system location is unavailable.

7.2.8 Scalability

In order to cope with future demands without too much capital expenditure up front, the network design needs to be scalable such that additional capacity can be added without a network redesign. You do not want to be in the scenario where you need to rip out the infrastructure being built when an additional 1000 end users are to be added to the service. Of course, no design can have infinite capacity without redesigning. However, it is advisable that you should keep the redesign to the existing network architecture to the minimum when additional capacity is required. For example, the original network architecture should be able to cater for service expansion to another geographic area or adding more network nodes. The additional nodes should present minimal disruption to the end users/customers that are currently on that service.

7.2.9 Network Management

Only effective network management can achieve customer satisfaction. The right attitude for providing a good service is to have a proactive approach. The network operators should notice and fix a network fault before the users and customers have realized it and log a fault ticket. One should strive to exceed the customers' SLA. This not only results in many happy users and customers, but the cost of running the fault management team will also be reduced. Chasing up a fault ticket and keeping the users and customers up to date with the fault situation is time consuming and costly. So how do we provide a good network management design?

The main object for network management is to have the facility for the network management center to monitor the network and services effectively. In order to do that we need to be able to:

- Collect information on what network elements are in the network. This is often known as network inventory (please see Section 7.5 for details). Without knowing what network elements are in the network, it is very difficult to manage, as you do not actually know what you are managing!
- Proactively monitor and collect information regarding the status of various network elements. In order to notice and fix the network faults before your users/customers complain, you need to have an effective way of monitoring the different network elements. Please see Section 8.11 for details on network management systems.
- Perform any remote diagnosis where possible. This will be the most effective way to find the fault if possible. Sending a man in a van is less than desirable. You only do it if you absolutely have to, as it is expensive.
- Collect information on the current network configuration and effectively manage the changes to the current configuration of all the network elements.
- Identify and isolate the cause of the faults as they occur and put in corrective action as soon as possible. This may involve rerouting network traffic around the failure node.
- Perform network performance analysis using the network utilization data available. Acting on the result of the analysis will help you to maintain the network at its most optimum performance. The performance data can also be used to detect potential faults early by spotting degradation of the service and will also provide information about whether additional network capacity is required.
- Perform network software upgrades and maintain the service configuration on each network element.
- Manage the security of the network elements and network nodes to prevent unauthorized intrusion. Network security will be discussed further in Section 7.6.

From the network design point of view, one needs to decide which network management protocol to use and ensure all the data required to support the above is being provided from the network elements. Otherwise, managing the network will become an impossible task. Most network elements use the simple network management protocol (SNMP) and common management information protocol (CMIP) to communicate the status of the network element to the network management systems. After the interpretation of this information, the operators at the network management center can take relevant actions to fix any fault where necessary. To manage the network successfully, a well set up network management system and a set of well thought out operational processes are required. Functions of network management systems and operational processes will be covered in Chapters 8 and 9.

7.2.10 Traffic Engineering

The objective for traffic engineering is to maximize network utilization and efficiency (and, therefore, the highest revenue potential at optimum cost) whilst delivering the required network performance and QoS to customers/end users. As part of traffic engineering:

- Traffic management functions are necessary to route network traffic to where network capacity is available but avoid inefficient use of network resources where possible. Much of traffic management functions are embedded as part of the network design. For example, routing and cost of routing, queuing and scheduling strategy to meet QoS requirements (as discussed in the previous sections of this chapter) are all mechanisms/engineering rules for traffic management.
- Performance analysis is performed to ensure that the network is efficient and meeting the performance requirements for the service and that any potential performance issues in the network are highlighted.
- A capacity management function is required to ensure sufficient network capacity is provided and that the suitable amount of network capacity is provided at the appropriate place to meet forecasted demand.

7.2.10.1 Requirements for Traffic Engineering

To achieve the above objective, one needs to know and understand:

- the source and destination of network traffic (i.e. the traffic flow through the network);
- the volume of traffic broken down by traffic type (i.e. the traffic mix by traffic type/QoS class);
- traffic pattern and profile by time of day;
- the network performance requirements;
- customers/end users contracted SLA or QoS;
- geographic forecast of network traffic;
- cost constraints.

7.2.10.2 Network Design/Traffic Engineering Rules

With all the input above, traffic engineering rules are to be devised as per the network design. These rules may include:

- routing policy and the cost of routing for each traffic type;
- overload control strategy;
- buffering/scheduling strategy;

- traffic queuing, shaping and discarding policy to satisfy the QoS requirements;
- bandwidth allocation per end user;
- channel selection and allocation of radio-frequency (RF) resources for 3G mobile network;
- QoS mapping between network domains for 3G mobile network;
- network resiliency and recovery strategy.

7.2.10.3 Ongoing Analysis and Monitoring

Analysis to be performed as part of traffic engineering includes:

- network performance monitoring and analysis;
- applying capacity analysis techniques to calculate/estimate capacity requirements;
- traffic modeling.

Further details on traffic engineering can be found in 'ITU E360.1 Framework for QoS routing and related traffic engineering methods for IP-, ATM- and TDM-based multi-service networks' [26].

7.2.10.4 Network Performance Analysis

The network performance requirements come from the service performance requirements of the service, the cost efficiency and network utilization requirements. When thinking about network performance analysis, one needs to ask:

- Is the network performance going to be measured end to end for the service?
- What are the network performance parameters to be measure and analyzed?
- What performance metric is most meaningful for the service?
- Which network performance parameters are realistic to measure without adding too much overhead in the network?

Service performance parameters/requirements that are related to the network can be, but are not limited to:

- call set-up and clear down time for voice services;
- session set-up and clear down time for PS-based services;
- end-to-end data throughput on uplink;
- end-to-end data throughput on downlink;
- end-to-end delay on uplink;
- end-to-end delay on downlink;
- end-to-end delay variation per application on uplink;
- end-to-end delay variation per application on uplink;
- MTBF and MTTR figures resulting in service availability.

Therefore, the network performance metric to be measured and monitored for a packet-based service may include:

- average throughput per customer/end user;
- average data transfer delay – delay measured from point A to point B;
- latency in the network – delay variation measured between point A and point B;
- average packet loss per network node;
- error rate per network node.

Other performance metrics (in addition to those listed above) for mobile networks/services might include:

- geographic coverage;
- call set-up time;
- lost calls percentage per cell group in regions;
- access failures (deny end-user access to network) per cell and per region/sector;
- bit error rate, frame error rate and signal quality estimate per cell;
- signal strength and transmission loss per cell and per region;
- radio blocking (due to congestion) per cell;
- usage and RF loss per cell;
- handoff failures per region.

Additional network performance parameters are also listed in Section 8.12.1.

7.2.10.5 Network Capacity Analysis

In general terms, network performance issues can normally be resolved by providing additional network capacity or modifying the network design. Network design has been covered in the previous sections of this chapter. In order to estimate the appropriate amount of capacity and avoid unnecessary expenditure, capacity analysis is required.

7.2.10.5.1 Voice Services
To perform capacity analysis and estimate the capacity requirements for voice/VoIP services, the following analysis/data will be needed:

- queuing analysis (e.g. Erlang theory to estimate the link and port capacity for voice switches);
- number of calls per busy hour;
- busy hour call attempts (BHCA);
- average call arrival rate;
- average call hold time;
- link occupancy.

7.2.10.5.2 PS Services
For traditional voice services, network resources/channels are dedicated to the end users once the call set-up has been completed. Therefore, estimation of capacity requirements is more straightforward than with those in the packet-switching environment, where most of the network resources are shared. The most important items for capacity analysis in the PS network are the volume of traffic, traffic profile and the traffic mix associated with each service/application. An asymmetric type application/service will have a huge impact on capacity requirements. For example, the capacity requirements for web browsing are very different from that of a VoIP service, where the former has a higher capacity requirement on the downlink toward the end user/customer, whilst the latter has equal capacity requirements on both uplinks and downlinks. Below are some network capacity metric examples per service/application you might want to consider:

- total number of data sessions per end user/customer;
- average throughput per end user/customer;
- total amount of data per downlink;
- total amount of data per uplink;

- average packet size per downlink;
- average packet size per uplink;
- average arrival times of packet sessions;
- percentage of transactions in traffic mix.

For capacity analysis for a PS network, please refer to *Designing and Developing Scalable IP Networks* [12] or *Designing Networks with Cisco* [57].

7.2.10.5.3 Mobile Services

Mobile services support both voice and data services; therefore, all the parameters above apply on a per cell site basis. In addition, the rate and number of call handoffs to another cell and any service/application-related traffic profiling (e.g. video conferencing service) should also be considered. Aggregated capacity from cell cites to the core network, both PS and CS domains, will also need to be estimated. Capacity and RF planning for mobile services are further discussed in Section 7.3.3.

7.2.10.6 Traffic Modeling

The principle of traffic modeling is to obtain a representative view on what is happening in the network in terms of traffic volume, traffic flow and network performance.

To devise a traffic model, one needs to consider:

- the traffic type and metric to be analyzed (i.e. taking input from the sections above);
- node sites to be included;
- size of each node;
- logical and physical network connectivities (including their bandwidth) between the node sites;
- routing/routing rules between the network nodes;
- network topology to be used;
- modeling techniques to be used.

Traffic modeling tools are widely available. Typically, traffic modeling tools can be used for network performance analysis and, to a large extent, capacity analysis (as both require a good understanding of the traffic pattern and traffic volume), as well as producing network trends. Traffic modeling tools can also be used to simulate potential traffic scenarios and failure conditions to predict network behavior under those conditions.

For further details on traffic modeling, please refer to *Traffic Engineering and QoS Optimization of Integrated Voice and Data Networks* [47]. Details on traffic engineering for mobile services can be found in *3G Wireless Networks* [7].

7.3 Service Network Design

Using the conceptual network architecture, as explained in Chapter 6, below are the different networks requiring design in order to support a service fully:

- access network;
- network termination devices/CPE;
- core network;
- signaling network;
- transmission network;

- OLO interfaces;
- network management network;
- service management systems network.

In this section, I have used three different example services (fixed broadband, VoIP and 3G mobile service) to illustrate any additional design considerations, as well as those mentioned in the previous section, for each of service network components (i.e. access network, CPE, core network, signaling network, transmission network and interconnections to OLOs). Since the network management network and service management systems network are not likely to be service or technology specific and are hopefully common across different network technologies, they will be explained in sections (Sections 7.3.4 and 7.3.5) that follow from the example services.

Please note that the examples given below are written with the assumption that readers have a basic understanding of the technologies concerned. References are made and further suggested readings are included for those requiring further details on the particular technology topics.

7.3.1 Broadband Services

An example of the network architecture for an ADSL service is shown in Figure 6.2.

7.3.1.1 Access Network Design

For broadband services today, there are a wide variety of broadband access technologies you can choose from. For fixed-line networks, broadband connectivity can be achieved using cable technologies or xDSL technologies over a copper pair or leased lines. As for wireless access technologies, Worldwide Interoperability for Microwave Access (WiMAX) and WiFi are now becoming popular. The choice of technology is heavily dependent on the bandwidth and mobility requirements.

The actual bandwidth a fixed-line end user can have is heavily dependent on the type of technology used in the access network, the quality of the copper line involved and the distance of the customer's/end-user's premises from the local exchange.

A leased line is normally chosen for users/customers that have a high bandwidth requirement from the core network to the customer site. In this example of a broadband service, leased lines will be used to deliver traffic from end users to data centers where traffic is terminated. Care needs to be taken for network resilience requirements, as the volume of traffic being delivered could be pretty high.

WiMAX (IEEE 802.16) and WiFi (IEEE 802.11) are becoming popular wireless access technologies. WiFi hot spots are normally used indoors, as the wireless signal range is relatively small (30–100 m), while WiMAX technology has a wider geographic reach (5–15 km).

7.3.1.2 Customer/End-User CPE Options

Customer/end-user CPE is dependent on the choice of access network technology. Quite often, the CPE is selected by the user themselves and the operators/service providers need to publish the interface and protocol standards supported by their network in order that the end user/customer CPEs is compatible with the network so that the service can be delivered.

7.3.1.3 Core Network Design

Traditionally, xDSL technologies are ATM based from the DSLAM to the core network. Therefore, when designing the network from the DSLAM onwards, it is essentially the same as designing an ATM network. The size of virtual circuits (VCs) dictates the maximum bandwidth end users can have in the core network. The QoS/CoS for the VCs is to be selected depending on the service/application on offer. ATM technology provides the following QoS categories:

- Constant bit rate (CBR). This is intended for nonburst real-time application, with a limit on cell delay variation (CDV) and cell transfer delay (CTD) specified.
- Variable bit rate (VBR). This has two sub-categories, namely VBR-real time (VBR-rt) and VBR-nonreal time (VBR-nrt), and is intended for bursty traffic, with cell loss ratio (CLR) target specified and CDV and CTD specified on VBR-rt.
- Available bit rate (ABR). This is intended for bursty traffic using some form of feedback-based rate control.
- Unspecified bit rate (UBR). This is intended for traffic that has no CDV, CTD or CLR requirements.

The CoS for the VCs also settles the CoS for the virtual paths (VPs). You also need to decide upon the size of the VCs from the DSLAM to the first ATM switch. The more VCs you can 'squeeze' into a VP, the higher utilization of network can be achieved. This, of course, cannot be applied to CBR VCs and VPs. For Internet-type bursty traffic, VBR-nrt is normally used. Further reading on ATM networks can be found in *ATM Networks Concepts and Protocols* [13].

Although the DSL services are traditionally ATM based, quite often the operators only use the ATM network as an aggregation network. The core delivery network can be IP based. This is because ATM ports are expensive and resource hungry for a broadband service, most traffic is IP-based connectionless traffic and it is more cost efficient to use a connectionless-based network (i.e. a IP network), even though some broadband applications may required QoS guarantees. New-generation DSLAM/multi-service access platforms are IP based.

If the core delivery network is IP, then IP network design considerations apply. Depending on the service being offered, certain QoS parameters might be required. For example, if video is to be supported as part of the broadband service, then the QoS requirement might dictate an MPLS solution. Care needs to be taken when mapping ATM CoS to QoS in the MPLS domain, as the queuing and shaping are different between the two technologies. If end-to-end QoS/performance guarantees are to be offered to customers/end users, then traffic queuing and shaping are applied to achieve the required QoS. When setting the QoS parameters, one needs to be very careful to avoid waste of network resources, which in turn will increase the cost of the service. Please refer to *Designing and Developing Scalable IP Networks* [12] for guidance.

In the example given in Figure 6.2, the interface between the ATM and IP network is the broadband access server. The broadband access server converts the ATM cells into IP packets and can offer other service features, like protocol filtering, traffic shaping, QoS policing and performance monitoring and so on. This is also where end-user service policies are applied.

7.3.1.4 Interconnection to OLOs

Quite often for a broadband service, the access network and core network are provided by different operators. Historically, the incumbent operators own and operate the access network. Hence, OLOs need to buy capacity from the incumbent. The network interfaces to these networks are very much dependent on the interfaces provided by the incumbent, and there is no single way of doing it.

Therefore, the interconnect network design is closely coupled with the interconnect network specification offered by the incumbent.

However, owing to deregulation of the telecommunications industry, many other operators can connect directly to the incumbent exchanges (i.e. LLU) and many operators are connecting to the copper wire or cable network directly. This may make the network design easier, as there is less dependency on other network providers.

7.3.1.5 Signaling Network Design

For data/packet switching networks in general, there is no separate signaling network. However, signaling protocols and channels are defined within various networks. For example, for an ATM network the default signaling channel has been allocated a VC identifier VCI = 5, which carries the signaling messages. For the IP networks, different signaling protocols or control messages are contained within the protocol used. For example, consider the session set up for a broadband end-user point-to-point protocol (PPP) session: the set-up request starts with the initiation of a challenge handshake authentication protocol (CHAP; RFC 1994) [63] request for user authentication requests. After successful authentication and authorization, the PPP (RFC 1661) [62] session can be established. All the control and request/ acknowledge messages are part of the PPP and CHAP control message within the protocol itself. There is no separate signaling network, the control/signaling messages within the respective protocols are used to communicate between different network devices. For further details on signaling and signaling networks, please refer to *Signaling in Telecommunication Networks* [59].

7.3.1.6 Transmission Network Design

The transmission network is the underlying transport for all traffic delivering the service. The most common technology used is SDH/SONET over wavelength division multiplexing (WDM) or DWDM fiber-optic networks. For the ATM part of the network, some operators may prefer ATM over DWDM directly. Please see the transmission design book for design considerations: *SDH/SONET Explained in Functional Models* [15].

7.3.2 VoIP Services (Fixed Line)

Figure 6.3 is an example of network architecture for a VoIP service inter-working with PSTN services. Below are some high-level network design considerations for VoIP services. For further details, please refer to *Voice Over IP Fundamentals* [19].

7.3.2.1 Access Network Design

The access network for the VoIP service can either be a LAN within an enterprise environment or a fixed broadband access network as described in Section 7.3.1.1. Depending on the number of VoIP and data connects that are required, there needs to be sufficient bandwidth on the LAN to support both VoIP and data traffic. Unless the CPE can mark and prioritize voice traffic before reaching the media gateway, it is advisable to 'overprovision' the bandwidth in the LAN environment to ensure the quality of the voice is not degraded before reaching the media gateway. For the broadband access network, this becomes less of a problem assuming the number of end users per media gateway/DSL or cable modem is limited.

7.3.2.2 Customer/End-User CPE Options

All CPE needs to be either H.323 or SIP enabled to support the VoIP service.

7.3.2.3 Core Network Design

There are three potential core network solutions for VoIP/packet voice application. The voice packets can be carried over an:

- ATM network;
- IP network or the Internet;
- MPLS network.

ATM is designed to support a multi-media, multi-service environment and has well-defined QoS parameters. Therefore, support for voice applications should not be a problem. However, ATM ports are relatively expensive and ATM networks are not as widely deployed as IP networks; therefore, it will be inefficient to transport IP voice packets over ATM. Therefore, ATM is not a preferred technology of packet voice traffic.

IP networks and the Internet are connectionless networks with little or no means to offer QoS guarantees. Since voice packets are sensitive to latency, jitter and drop packets and so on, the quality of the voice at the receiving end may not be comprehendible without these issues under control. Therefore, an IP network or the Internet is not a preferred solution for quality voice, without appropriate labeling or QoS rules being applied – see below. However, it is a cheap solution.

MPLS protocols enable network traffic to be forwarded or switched using a simple label (as detailed in RFC 3031 [16]). It can run over IP, ATM, frame relay, Ethernet and IP networks. MPLS technology adds a label (according to the destination IP address, host address, and host address and QoS) to the data packet and forwards it to the next hop.

The capability to support QoS and CoS can be achieved through the use of DiffServ (e.g. DSCP) and IntServ (e.g. RSVP) techniques, and the ability to perform traffic shaping and engineering. MPLS, together with the use of DiffServ, is a favorable solution to the packet voice service.

Within the MPLS solution, there are two ways to carry the voice packets: encapsulating using IP or voice carried directly over MPLS without IP encapsulation. Carrying voice directly over MPLS is, of course, more efficient; however, it will be a challenge to have MPLS deployed in the access network. The most common packet voice solution is VoIP with an MPLS core network. Further details on QoS using MPLS can be found in *Advanced QoS for Multi-Service IP/MPLS Networks* [35] and *Voice over MPLS: Planning and Design Networks*.

7.3.2.4 Signaling Network Design

Different signaling protocols have been developed to address the need for real-time session signaling over a packet-based network. H.323 was initially developed in the enterprise LAN community, whilst SIP was developed by the IETF. This scenario arises due to the nature of packet-based networks, where, historically, only nonreal-time traffic is carried. Therefore, both SIP and H.323 are the signaling protocols used for VoIP. Signaling is exchanged between the media gateways, media gateway controllers and signaling gateways, sometimes known as trunk gateways (as seen in Figure 6.3).

Signaling transport (SIGTRAN) protocols specify the means by which SS7 messages are reliably transported over an IP network. SIGTRAN facilitates the inter-working of VoIP networks with PSTN, thus making any telephone in the world reachable from either network.

There are no separate signaling network/dedicated channels for VoIP. The signaling within the IP core network depends on the QoS implementation. For IntServ, RSVP is the signaling and control protocol for requesting and reserving network resources to achieve the QoS level required for voice application. For DiffServ, DSCP is the technique used to provide different levels of service for voice packets.

7.3.2.5 Interconnection to OLOs

The major area for interconnecting to other operators for a VoIP service will be the voice breakout to the PSTN network. You can interconnect to other operators' data network for voice application. As mentioned in Section 7.3.2.4, signaling plays a major role in inter-working of the PS network and the PSTN network. Design considerations for these types of connection include:

- the compatibility of standards of the VoIP signaling gateway deployed and the SCP being connected to;
- the resiliency and reliability of the network delivering the VoIP signals to the SCP;
- various failure scenarios on either side of the network;
- the end-to-end QoS for the voice packet, especially if you are delivering PSTN voice to the data network;
- the compatibility of features offered on the VoIP signaling gateway as opposed to that of the SCP.

For connection to another operator's data networks, the area of major concern is the transparency of QoS and CoS between the two data networks. Depending on the network technologies deployed by both parties, the end-to-end QoS for the voice packets may not be achieved. Even if both parties are using MPLS IP networks, the CoS definition may still vary.

7.3.2.6 Transmission Network Design

The transmission network is the underlying transport for all traffic for the service. If it is a LAN infrastructure, then the underlying transmission technology will probably be Ethernet. If an xDSL network is used in the access network, then the transmission technology will be DSL.

The most commonly used transmission technology for the core network is SDH/SONET over WDM or DWDM networks. The most common network topology of an SDH network is ring structured with diverse network path protections. For transmission network design, please refer to *SDH/SONET Explained in Functional Models* [15].

7.3.3 3G Mobile Services

Factors to be considered when designing 3G mobile services from a network perspective is described in the section below. Figure 6.4 gave an example network architecture for a 3G mobile service.

7.3.3.1 Customer/User Handsets/Network-Terminating Devices Options

For mobile handsets and network-termination devices, a list of compatible handsets for different network operators is available in most mobile-phone shops. Since there are different 'air interface' standards available (e.g. CDMA, WCDMA), all mobile devices are tested for compatibilities in the operator's network before releasing to the customers/end users. The handset functionalities and capabilities often provide or restrict the application/services that the end users can use and the service providers can design.

Another factor is the pricing of the handsets. This is directly related to the amount of subsidies the operators are giving the end users/customers for each type of handset. The commercial arrangements between the operators and the handset suppliers may vary between different handset models. The operators will only release those handset/network-termination devices that make commercial sense for them.

7.3.3.2 Access Network Design

As seen in Figure 6.4, the access network for a 3G mobile service is essentially the air interface between the UE and the core network. It connects the user handsets and the core network. The elements that need to be designed are the RF network and the transmission links from the BSS/RNS to the core network.

7.3.3.2.1 Access Network Requirements
Before we can design the access network, we need to know:

- The type of end users' equipment needs to be supported. However, compatibility requirements between the handset and network could work both ways.
- The geographic area coverage required and what bandwidth is available to end users within each of the geographic areas.
- The spectrum available for use.
- The spectrum required to achieve the service bandwidth requirement.
- The forward and backward compatibility requirements on the access network standards to be used.
- QoS and grade of service (GoS) requirements for the service and traffic types and profile.
- What 'air interface' standard is being used? CDMA or WCDMA?

7.3.3.2.2 Access Network Elements
Elements within the access network include:

- BTS/base station (BS) to end-user handsets/terminals;
- BTS to BS controller (BSC)/BS to radio network controller (RNC);
- BSC/RNC (GSM/EDGE RAN (GERAN)/UMTS RAN (UTRAN); GSM = Global System for Mobile Communication, EDGE = Enhance Data for Global/GSM Evolution) to PS core network;
- BSC/RNC (GERAN/UTRAN) to CS core network.

7.3.3.2.3 RF Coverage and Capacity Analysis Between the Handset and BTS/BS
When you ask the marketing department what geographic coverage they need, the answer is, of course, everywhere! This is obviously not very useful to any network designers or network planners. In order to plan and design the RF network (air interface) effectively, we need to understand which geographic areas are urban and suburban, what areas are primarily commercial, residential and industrial, which areas are parkland, where are the motorways located, and what is the population in each of the areas and the service penetration within these areas.

In addition, in a given area, we need to know which types of service are available (e.g. speech-only service available?) and what data rate is required for each end-user handset/terminal. This information is important, as the effective footprint of each cell is highly influenced by the data rate to be supported. The higher the throughput rate, the smaller it is for the effective cell radius.

Another important input is the link budget. This is the calculation of the amount of power received at a given receiver based on the output power from a given transmitter. The link budget accounts for all the gains and losses that the signal can be 'heard' at the receiver. The maximum available path loss in both directions (uplink and downlink) needs to be determined to find the coverage limit for the cell for the service on offer.

A further design consideration in the RF network is the inter-cell interference. The higher the interference from nearby cells, the lower the capacity; hence, only a small footprint can be achieved for a particular cell. The lower the interferences from nearby cells, the higher the capacity; hence, a bigger footprint can be achieved by a particular cell. In effect, one cell can 'borrow' capacity from another cell that is less loaded.

Other factors to consider for a mobile access network that will affect the network performance include:

- attenuation;
- free space loss;
- Doppler shifts;
- reflection, multiple paths and delayed spread.

When we have a good understanding of the coverage requirement, the link budget and the inter-cell interference, we can then start generating the RF coverage plan. Since RF planning is a huge topic and a book can be written on this subject alone, I do not intend to go into detail here. Details on RF design and capacity analysis can be found in *3G Wireless Networks* [7] and *Design and Performance of 3G Wireless Networks and Wireless LANs* [34].

7.3.3.2.4 Access Network Transmission Design Between BTS to BSC/BS to RNC

One of the first things we need to know is the number of BSCs/RNCs that are required. For many GSM BSCs, the main capacity limitation is the number of BSs/cells that can be supported. With a UMTS network, the capacity of most RNCs is tightly linked with traffic mix, the total traffic throughput, the number of BHCAs for voice traffic, and the total Iub interface (physical transmission interface between base stations and RNCs) capacity into RNCs.

Other things that need to be decided may include the:

- minimum number of RNCs (BS) needed, including the capability for future growth and 'soft' handover to minimize the total switching demands on the RNCs;
- transmission capacity and network design.

The RNC network design is also closely linked with RF design for coverage and capacity demands. One needs to be aware that the RF network is the greatest part of the total network cost. Therefore, deciding the number of RNCs can have a great effect on the total capital expenditure for the network.

7.3.3.2.5 UTRAN/GERAN Transmission Network to Core Network

When designing the UTRAN/GERAN transmission network, we need to consider:

- how much inter-RNCs/BSCs handover traffic there is;
- the user traffic throughput.

The network topology can be ring or mash, depending on the cost and distance between nodes. As per the UMTS architecture, BSCs, RNCs, service GPRS support nodes (SGSNs; GPRS = general packet radio service) and mobile switching centers (MSCs) all have ATM interfaces; therefore, traffic between these nodes is ATM based. ATM network design considerations can be found in *ATM Networks Concepts and Protocols* [13].

7.3.3.3 Core Network Design

Within the UMTS framework, the functions performed within the core network include:

- authentication of end-users' identity and authenticity;
- identification of end-users' equipment, ensuring the handset is genuine;
- call control, involving voice calls establishment and release between end users and other connection networks;
- session management for data session, relating to the establishment and the release of data sessions between end users and destinations of data packets/end points;
- management of QoS requirement for both voice and data session where relevant;
- switching for voice calls to establish and release voice connections;
- routing for data packets to be transferred and delivered to their destination addresses.

The UMTS core network consists of (as illustrated in Figure 6.4):

- the CS domain;
- the PS domain;
- entities shared between all the domains.

7.3.3.3.1 CS Domain

The CS domain provides network connectivity for voice-related services. The visitor location register (VLR) obtains end-user profiles from the home location register (HLR) and the MSC provides the switching functions based on that information. The gateway mobile switching center (GMSC) provides the connectivity and interfaces to the external CS networks. The CS network carries the conventional voice traffic and provides connectivity to other PSTN service providers.

7.3.3.3.2 PS Domain

In the PS domain, the SGSN obtains the end-user's profile and service information from the HLR and provides routing functions for end-user data packets. The PS network carries network traffic for the data network services and provides Internet connectivity where relevant. The gateway GPRS support node (GGSN) provides the onward connectivity to external packet-switching networks.

Design considerations for the PS and CS parts of the core network include:

- network topology and architecture;
- network node sizing;
- routing strategy;
- scalability;
- interconnections to OLOs and signaling standards used.

Design considerations described in Section 7.2 are relevant for both CS and PS network designs. Further details on design considerations for PS (IP) networks can be found in *Designing and Developing Scalable IP Networks* [12].

7.3.3.3.3 Entities Common to All Domains

The entities common to all domains are:

- HLR. Provides end users location registration and management support functionalities and provides call/session establishment support.
- Authentication center (AuC). Provides end-user identification, end user's security information, end user's access authorization and service authorization support functionalities.

The above are grouped together as part of the home subscriber server (HSS). HSS functions also include providing service profile data to CS, PS, IP multimedia subsystem (IMS; please see below), application services and other service-specific entities, as well as service provisioning functions:

- Equipment identity register (EIR). Stores the end user's equipment details and status of this equipment. Any suspect ('gray-listed') equipment details are stored and calls are delivered but may be traced. No traffic is delivered to and from any stolen/'black list' equipment.
- IMS.

IMS has been created to include the functionalities above plus the SIP interfaces and functions for call/session controls and management, IP policy control, service profiling and service functions with interface to the application servers.

Further details of IMS architecture and functionalities can be found in *The 3G IP Multimedia Subsystem (IMS): Merging the Internet and the Cellular Worlds* [45] and *UMTS Networks: Architecture, Mobility and Services* [46].

Within the UMTS service, other entities may also be present to provide other services (e.g. SMS gateways and switching centers to provide SMS).

7.3.3.3.4 Dimensioning of Core Network Nodes

When dimensioning the core network nodes, the following factors should be considered:

- total number of end users;
- BHCAs (for voice services);
- maximum Erlang to be supported (for voice services);
- number of simultaneous session requests;
- total number of active end users at any one time;
- total number of roaming end users to be supported;
- total number of end-user roaming sessions to be supported at any one time;
- throughout per end user;
- total number of RNCs and BSCs to be supported;
- the ratio of number of MSCs and SGSNs to RNCs and BSCs.

7.3.3.4 Addresses and Identifiers

Unlike a fixed network service, you need to be aware that UMTS networks require many kinds of numbers, identifiers and permanent and temporary addresses to uniquely identify the end users, enable signaling functions, perform service separation, location identification, provide security protection and call and session routing. Please refer to *3G Mobile Networks Architecture, Protocols and Procedures* [5] for details.

7.3.3.5 IP Addressing and Routing

For a 3G mobile service, the mobile IP framework can be used to maintain network connectivity when the mobile terminals are on the move. The mobile terminals, known as mobile nodes (MNs) in mobile IP terms, have home (IP) addresses assigned. The home addresses are used within the home network to route the addressed datagrams to and from MNs and are used to communicate with the home agent (HA) when in foreign networks. (The HA is a router on an MN's home network to maintain current location information for the MN and tunnels the addressed datagrams to the MN when located away from the home network.) Within the home network, the conventional IP routing considerations apply.

When the mobile terminals are in a foreign network (i.e. roaming to a different network), care of addresses (CoAs) are assigned. For IPv4, the CoAs are assigned by the foreign agent (FA). (The FA is a router in the visited network that the MNs are associated with to indicate its location.) After FA registering (the CoAs of MNs with their respective HAs), datagrams addressed to the MNs are tunneled from the HAs to FAs for delivery to the MNs. Hence, the FA provides routing services to an MN and decapsulates tunneled datagrams by the HA and delivers them to the MN. The FA may also act as the default router for datagrams sent by an MN when visiting other networks.

There is no requirement for an FA while in IPv6. MNs are always addressable by their home address. The MN can configure the CoA by stateless address autoconfiguration or neighbor discovery. The association between the MN's home address and CoA (known as bindings) is maintained by the corresponding nodes (CNs) that the MNs are associated with at the time. (A CN is a peering node that the MN is communicating with.) This binding is registered with the HA and retained by the HA as a cache entry. The IPv6 encapsulation is used to communicate between the MNs and HAs. Data traffic can be exchanged between MN and HA via tunneling (and reverse tunneling). The MN and CN can exchange data packets without the support of HA. The CN sets the destination address within the IPv6 header to the CoA. The home address is in the new IPv6 'home address' destination option.

For full mobile IP, registration, routing considerations and operations, please refer to *Mobile IP Technology and Applications* [8] or *Mobile IP Technology for M-Business* [33].

7.3.3.6 QoS Considerations

Within the UMTS framework, different QoS traffic classes are defined as:

- Conversational class. This is most stringent of all classes, with strict upper bounds on transfer delay and delay variation, and no requirement bit error. Typical applications for this class are voice and video conferencing.
- Streaming class. This is characterized by the real-time one-way data flow with constraints on delay and delay variation requirements and no requirement on bit error. Typical applications in this class include streaming videos and video-on-demand services.
- Interactive class. Best described as a request-and-response-type operation, where a request is followed by a download of data from the request. This traffic class has high tolerance to delay; there is no requirement on delay variation, but there is a stringent bit error rate. Typical applications for this traffic class would be web browsing and downloading files/information.
- Background class. This has no requirements on delay or delay variation, but with a stringent bit error rate. The applications using this traffic class are of low priority and can run in the background (e.g. e-mail and SMS).

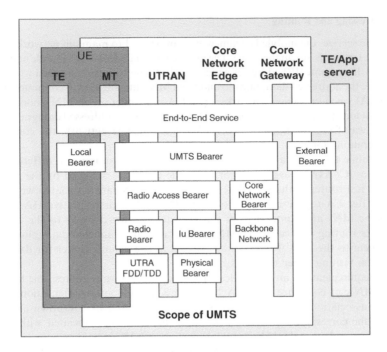

Figure 7.1 QoS architecture in UMTS network.

The QoS architecture in the UMTS framework is illustrated in Figure 7.1. The end-to-end service is defined to be between the mobile terminal of the UE; that is, the mobile handset/terminal and another mobile terminal or an application server (depending on the service scenario). However, the scope of the UMTS bearer service runs between the mobile terminal and core network gateway.

To achieve end-to-end QoS within the UMTS environment, performance characteristics within each traffic class are to be negotiated for each end-user session and these characteristics are maintained throughout the whole service path, from source (the end-user's mobile terminal) to the destination, where this could be other mobile terminals or servers/devices within the terminating network. Mechanisms to achieve this include the use of appropriate control signals, appropriate QoS guarantees within the radio access bearer service (e.g. radio resource management strategies) and the core network bearer services, and a consistent approach with QoS management functionalities within and between each of the bearer services. A consistent management of network resources when connecting with an external network also plays an important part in ensuring end-to-end QoS. The translation/mapping function in IMS can provide the conversion between the QoS requirements by the application/IP QoS parameters and UMTS QoS parameters. Further discussions on this can be found in *End-to-end Quality of Service over Cellular Networks* [44] and *IP in Mobile Networks* [50].

7.3.3.7 Network Interfaces to Other Network Providers

As mentioned previously, the GMSC provides the connectivity and interfaces to the external CS networks. The CS network interface carries the conventional voice traffic to and from other PSTN service providers.

For a PS network, the interface is provided by the GGSN. The PS network interface provides the onward connectivity to external packet switching networks and receiving data packets (e.g. tunneled traffic for visiting end users from their home network).

7.3.3.8 Signaling Network Design

For a UMTS service, there is no separate network for signaling. The signaling control functions for session or call set-up and so on are performed within the relevant protocol. Unlike the conventional PSTN world, where the signaling and control function is via the SS7 signaling network, the control functions for call set-up within the UMTS framework are done by the SIP (RFC 3261) [65]. Any bandwidth reservation functions may be performed by the RSVP, whereas the session description protocol (RFC 2327) [66] may be used to describe the multimedia session or RTP (RFC 3550) [67] and real-time transport control protocol [68] are used for transporting real-time data and providing QoS feedback.

7.3.3.9 Transmission Network Design

The transmission medium for the access network is radio frequency and the technology used is dependent on the standard being deployed (i.e. are you using CDMA or WCDMA or other proprietary protocols), whilst the rest of the network can be SDH or other transmission technology (e.g. microware). The most common technologies for the transmission network are SDH or SDH with DWDM.

The network topology for the transmission network for the access network is normally a tree-and-branch topology. While in the core, the ring topology with a logical mash is often used. Further design considerations on transmission network design can be found in *SDH/SONET Explained in Functional Models* [15].

In a 3G mobile network, the transmission is often not considered a priority, as most effort is spent on optimizing the RF network. However, upgrading and building a transmission network involve not only the operators laying the fibers, but also other complex issues like obtaining permission to lay the fibers where required. This should not be overlooked, as this might take time to resolve.

7.3.3.10 Traffic Forecasts and Capacity Planning

As with all services, the basic forecasting information required would be the number of end users and customers for the service and the growth-rate timeframe. In addition, as mentioned in Section 7.2.3, the traffic forecast for each of the service types is a major input into the network design activity. For 3G services, this is more complicated, as end users are continually on the move and a combination of services involving voice only, voice and data or data only services can be active simultaneously.

For voice traffic, the forecast will need to include the distribution of traffic in terms of mobile to land, mobile to mobile and land to mobile. Since the mobile and land traffic are delivered to and obtained from different networks, such details will be essential. In addition, for the land traffic, a percentage differentiation between local and national (long-distance) traffic will be useful. As per forecast of any voice service, the average number of calls per user in the busy hour and mean hold time per call is a must have for any voice traffic forecast. Using Erlang B theory, we can derive the number of BHCAs to size the voice network accordingly.

For data services, there is also a distinction between user/subscriber traffic and commercial/ customer traffic, as their profiles are very different. In addition, the traffic mix of different types of data traffic is also very important, as the throughputs of different traffic types are very different. For example, the throughput profile of web browsing typically requires a high demand of bandwidth on the downlink, whereas the imbalance of throughput demands between the downlinks and uplinks is less for e-mail traffic. The traffic profile also determines the throughput of the source and destination of the traffic. For example, for an operator providing e-mail and an Internet service, e-mail traffic will be terminated at the e-mail server within the operator's own network, whilst the Internet traffic will be delivered to the Internet via the Internet traffic peering points. Both network elements will have to be sized differently according to the traffic profile.

Since the voice and data traffic are handled by different network elements and systems within the core network, the traffic profile of the two different traffic types will have a great bearing on the network node sizing of the core network. For example, the voice traffic in 3GPP release 1999 is handled by the mobile switching center (MSC) and data traffic by the packet data service node (PDSN). Therefore, the dimensioning of an SGSN or PDSN can be independent of each other and the BHCAs for voice traffic will be of no consequence in the core network even if the peak data traffic occurs at the same time. In the access network and the transmission network, both voice and data traffic are carried over the infrastructure, so the voice and data busy hours need to be considered together.

Ideally, a monthly traffic forecast should be provided. However, all these forecasts are estimates projected from the combination of number of users/customers and their existing usage patterns. These figures should be used as a guide rather than as absolute. Therefore, continual traffic monitoring for the traffic profile after service launch is very important and the network design should be scalable to cope with unforecasted demands.

Unless you have a magic wand, the network does not just appear overnight when there is a sudden increase in demand. Therefore, the network operator needs to plan and build the network ahead of time according to the forecast. Typically, at the start of the service, the operator will build the network capacity up to the first 12 months of the service forecast. Then, the network build will reduce according to service take-up. The network capacity and build activity is, of course, largely dependent in the CAPEX spend available. Network and capacity planning of a service will be covered in Section 7.5.

Guidance on network dimensioning and factors to consider when designing and building the network to support 3G mobile services can be found in *3G Wireless Networks* [7]. For capacity analysis for mobile networks, please refer to *Design and Performance of 3G Wireless Networks and Wireless LANs* [34].

7.3.4 Network Management Network Design

As seen in Section 7.2.8, the network needs to be managed through network management systems. To connect the network elements to the management system, a network management network needs to be designed. In most operator/service provider environments, a network management network should already exist. The questions will be:

- What is the network management architecture to be used?
- How many network devices are to be monitored and how frequent will they polled?
- What is the average number of interfaces that require monitoring per network element?
- What are the network parameters that require monitoring and measuring?
- How much additional network management traffic will this new service generate?

- Does the current network management network have the capacity to cope with the additional network management traffic generated by the new network elements?
- Are there any performance requirements on the network management traffic?
- Is the existing network management system being used?
- Where does the network management traffic need to go to and from?
- How will the new network elements be connected to the existing management systems?
- Do you design and build a new network management infrastructure?
- At what point will a new network management network be required?
- If existing infrastructure is used, how do you ensure the traps from this service are not mixed with the others?

Most network elements have a separate interface (normally Ethernet interfaces) for network management purposes. These are connected to the network management network such that the network elements can be managed by the network management system.

If a new network management system is required for various reasons (e.g. the lack of capacity on the existing system or the network view/scheme is drastically different from the current network), then potentially a new network management network design is to be created.

7.3.5 Service Management Systems Network Design

There are two types of service management systems: those that face the customers (e.g. customer reporting system, orders capturing and management system and billing system) and those that face the network (e.g. the system that provisions the customers and end users on the network, the systems that manage the users' policy and profiles; the systems that collect data from the network to produce bills and reports).

The network linking these systems to the network and the network enabling the systems to communicate with each other are often forgotten about until too late. This is because the systems designer takes the network for granted and the network design does not really care about service systems. The system designers often say, 'It's only a piece of wire linking the systems, isn't it?' How many times have you heard the network designer saying, 'Oh, it is the IT department's problem, not mine! I don't touch those!'? Hence, this item often falls between the gaps. To fix this problem, the service designers need to ensure that the ownership of this piece of work has been assigned and that the design will be able to support the service.

For the systems facing the network, there need to be communication links between these systems and the network elements (e.g. provisioning system and the system holding the user's policy and profile). If policy and profiles of the end users are applied as part of call/session set-up, then there may be a need for dedicated and resilient links from the network elements concerned to the end-users' policy system. Loss of revenue may result due to link failures. In this scenario, it might be wise to co-locate that system with the network elements concerned to reduce latency on the network link.

In most operator or service provider environments, there are internal networks linking different systems together. One must not forget to check if the current internal network infrastructure can support extra traffic that is being generated by the introduction of the new service.

Depending on the service requirements, there might be a need to establish secure network connections for the customers/users to access the customer-facing systems. For example, the service might include a requirement where customers can download the bill securely via the Internet. Then, both the system and network to access the system need to be designed to cater for this requirement.

7.4 Network Security

Network security is a topic of its own right and I have no intention in covering it in detail here. However, from a service design point of view, one needs to recognize the threats and ensure that risks are minimized.

For the voice network, the most common security threat is eavesdropping. This can be done by means of bugging, wire tapping and so on. Therefore, physical network security needs to be maintained throughout the physical network infrastructure.

From a data services point of view, there may be requirements for the network to provide a mechanism for application-layer security (e.g. SSL or secure hypertext transport protocol (SHTTP)) or different forms of encryption (e.g. single key or public/private key). However, these requirements do not necessarily protect the network from security threats. Figure 7.2 highlights some of the most common security threats.

The objectives for providing a secure network environment are:

- Providing a way to identify and authorize the users – authentication.
- Ensuring the authorized users have the appropriate access to resources they are authorized to use – authorization.
- Preventing sensitive data from unauthorized use or access – confidentiality.
- Ensuring the integrity of the data being transmitted (i.e. the content of the data packets are not altered in an unauthorized manner) – integrity.
- Preventing malicious damage to hardware, software and facilities – access security.
- Preventing accidental damage to hardware, software and facilities.
- Communicating to each employee to respect and maintain the information security policy of the company.

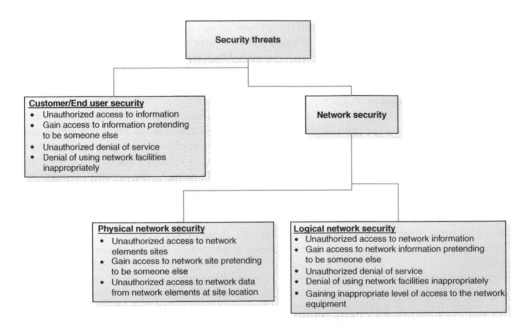

Figure 7.2   Security threats.

Each company should maintain a security policy to minimize the risk of security treats and to protect sensitive company data. ISO/IEC 27002:2005 [29] is a good standard to adopt with regard to security policies and procedures.

7.4.1 The Support of Customers' and End-Users' Security

Authentication is the way to identify that the user/customer is who they claim to be. The common method of authentication is to use a username (or user ID) and passwords to verify a customer's/ user's identity. Hackers often 'sniff' the user IDs and passwords of the end users. To avoid passwords being 'stolen', token-based password schemes are often used. As the password code changes every minute, it makes the hacker's life much more difficult.

Authorization controls the access level that each end user has. Therefore, after authentication, the end users/customers are only authorized to use certain parts of the network or service being provided. The most common protocol for authentication and authorization is RADIUS (RFC 2865) [11].

Denial of service is a very common malicious attack. In terms of the end user's/customer's network, this is caused by a user sending a huge data packet to another end user's/customer's router/network-termination device. This can be dealt with by configuring the router or firewall to block data packets over a certain size. Alternatively, using dynamic address allocation during session start-up can also reduce the risk of being attacked.

Encryption is a good way to ensure end-to-end data transmission confidentiality of user's/ customer's traffic. Both the sender and the receiver must have a compatible encryption and decryption procedure such that the data is meaningful to both sides. Other protocols that can protect end-to-end transmission security include Internet protocol security (IPSec), SSL and SHTTP/HTTPS.

7.4.2 Physical Network Security

To prevent unauthorized entry to network nodes (including external cables and ducting), appropriate security measures are required at the node sites. Locking the nodes sites and putting up wired fences are the simplest way to guard the node sites. For a larger site, you might want to consider implementing a security system that uses security passes with different levels of access rights to different parts of the network node site or building for different types of personnel.

7.4.3 Logical Network Security

Authentication and authorization also applies to operational staff that can gain access to the network elements. Different levels of network security access/authorization should be implemented. For example, only certain personnel can change network configurations or disable network services. Others can only have read access of the network elements. This is a good way to reduce unauthorized access to network information and reduce the risk of either accidentally or intentionally causing damage to the network. With the appropriate levels of authorization in place, the network facility can only be used by the appropriate personnel. An access control list can also be used to restrict access to network elements only from listed IP addresses.

Spoofing can be used to gain access to the network by pretending to be someone else or the data packets appear to be from a source that they are not. A good way to protect against these types of attack is to implement source filtering at the routers or firewalls. Only data packets from certain sources are allowed to access the network elements. SSL and SHTTP are also good ways to prevent spoofing.

7.4.4 3G Mobile Security

The access network within the mobile network infrastructure is the most vulnerable to security attacks. Therefore, different techniques have been employed in the 3GPP standard to ensure that data confidentially and integrity is kept and that end-users' privacy is protected. I have highlighted some additional security mechanisms deployed below.

7.4.4.1 Mobile End-User Security

To prevent the unauthorized use of mobile terminals, the mobile end users should implement password protection on their mobile terminals' USIM (just like logging onto the PC or work station). The UMTS server domain security framework can only authenticate the mobile terminals and there is an inherent assumption that the mobile end user using the terminal is authorized to do so.

In addition to encryption, mobile end-user sessions are further protected using replay protection. This prevents session interruptions from hackers impersonating the mobile end users by sending a valid packet after a certain time. Replay protection employs a sequence of numbers and/or time-varying parameters to ensure the data packets sent/received are current.

Other security measures concerning the mobile end users include:

- end-user identity confidentiality (please see below for details);
- end-user location confidentiality – the locations of end users are kept confidential;
- end-user traceability – ensuring the service delivered to the end users cannot be traced.

7.4.4.2 UMTS Network Access Security

The confidentiality of mobile end-users' permanent user identities (international mobile subscriber identity (IMSI)) is protected by using temporary identities to address the mobile end users when transporting data and calls once the identity of the mobile end users has been authenticated. Whilst the mobile equipment identification is checked against the data held in the EIR for legitimacy.

Mutual authentication is used for authentication of the end users and the network. This ensures both that the users' identities are authenticated and that the network is authorized by the end-users' home networks to provide service. This is performed between the end users' mobile terminals, the home networks and the servicing networks. The mechanism used within the UMTS environment to provide mutual authentication is the 'authentication and key agreement'. It is a challenge and response technique where both the challenges and the responses are encrypted and have a shared 'master key' between the end-user's mobile terminal and the home network where encryption and integrity checks are derived. Please refer to *UMTS Networks: Architecture, Mobility and Services* [46] for further details.

To prevent eavesdropping of end-user traffic, the communication between the mobile terminals and the UTRAN is encrypted. As part of the integrity protection, the individual control signals between the mobile terminals and the RNC (within the UTRAN) are also individually authenticated.

7.4.4.3 UMTS Network Domain Security

The network-layer security within the UMTS network (i.e. between all the core network elements, known as network domain security (NDS), is provided by the IPSec protocol suite (RFCs 2401–2412) [60]. IPSec is used to provide secure communications between the network entities at the IP layer.

MAPSec (mobile application part security) is used to protect the application-layer security and the SS7-based network elements within the UMTS network. It secures the communication between the two network entities from the application-layer signaling point of view, rather than protecting the user data transfer for the application.

The security between network domains (e.g. between mobile service providers or operators) is provided by a security gateway (SEG). SEGs use IPSec to communicate with each other, and the security associations are set up using internet key exchange (IKE) [RFC 2409].

Further details on 3G mobile security can be found in *3G Mobile Networks Architecture, Protocols and Procedures* [5] and *UMTS Networks: Architecture, Mobility and Services* [46].

7.5 Network Inventory

As mentioned in Section 7.2.9, in order to provide effective network management for a good service, it is essential to know what network elements need to be managed. One of the mechanisms to provide that view is through a network inventory database. When building and implementing the service, it is vital to ensure that all equipment (including ports and the available capacity) is logged in the network inventory database, otherwise management of this new equipment will become impossible.

The network inventory can also be used to monitor capacity updates, assess network capacity availability on certain network elements, capacity reservation for service provisioning, fault management and service impact analysis during fault conditions. Details on data to be held as part of the network inventory database can be found in Section 8.11.3.

7.6 Capacity Planning, Network Planning and Optimization

As demand for the service increases, the requirement for additional network capacity will become apparent. Network capacity does not appear overnight. On the contrary, the typical lead time is 6–9 months; hence, the need for capacity planning and network planning arises.

Capacity and network performance analysis (as described in Section 7.2.10) are performed to reveal where additional capacity is required and if overcapacity provision exists. If possible, you might want to reroute some of the traffic such that the network load is more balanced. Where additional network capacity is unavoidable, network planning activities need to take place to build additional capacity.

In order to plan the network capacity effectively, we need to have

- a traffic forecast for the service over the network;
- traffic type;
- an indication of source and destination of traffic;
- the population and type of usage (i.e. residential or commercial) per geographic area.

With all the input data above, performance analysis and experience with the network and service usages data, we can then plan where additional capacity is required. Depending on the network build lead time, ideally it will be good to have 'just in time' provisioning of network capacity to reduce capital spend as much as possible.

7.6.1 Network Optimization

The objectives of network optimization are to:

- optimize the network for operational efficiency;
- reduce capital spend by 'sweating out' the existing network asset and reusing the current infrastructure to improve network utilization;
- remove network bottlenecks to improve network performance and end-to-end dimensioning.

This might involve rearranging network capacity, rerouting some of the traffic or the reuse of same network capacity with traffic of different usage pattern. As a simple example, business traffic normally utilizes the network during office hours (09:00–18:00) and busy hours for residential Internet traffic is normally in the evening (between 16:00 and 01:00). With these different usage patterns, you can engineer your network such that both types of traffic share the same network asset without building additional capacity. Network optimization is a topic in its own right. Further details on network planning and optimization can be found in *Mission-Critical Network Planning* [48] for fixed networks and *Fundamentals of Cellular Network Planning and Optimisation* [49] for mobile networks

7.7 Service Configuration in Network Elements

In order to achieve high network utilization and cost efficiency, one network element often supports multiple services. For example, an MPLS edge router may support VPN, Internet, voice and video services. Therefore, the ability to manage these different configurations on the same network element without jeopardizing the different services is becoming more complex, but vitally important. In addition, there should be minimal impact (e.g. performance degradation) on the current services when new ones are being added to the same network elements. These service configurations on the network elements will require management and control. As the demand for services to be supported by the same network grows and the services are more sophisticated and ever more changeable, it will become impractical to hold and manage the service information and configuration in the network elements.

In the traditional switched circuit world, the first creation of service-based controlled policy platform is the intelligent network, where information and call routing of the service is held in separate platforms (as opposed to embedding the details in the circuit switches). With services going towards personalization, it is becoming impractical to have all the customers/end users and their services and service profiles held as part of the network. This trend and the need to separate service and network functions (as mentioned above) forces the session management, service profiling and service/end-user control policies to be held outside the network elements in another central location, usually in a different physical location from the network elements. The advantages of centralized control include:

- It is much easier and, hence, more cost effective to manage all service profiles and control policies from one place. Service provisioning and changes to the service profile and control policies only need to be made in one place.
- Most network elements only service a certain geographic area. If the end user moves to another location, then any updates are only made centrally. This is especially true for mobile services, where the service profiles for the end users are accessed from different locations all the time.

There will be fewer interfaces to be managed for the same data held; hence, there will be improved operational efficiency.

With a centralized service control platform, separating the service configuration from the network elements, multiple service/application providers can have easier access to the network via the service platform. Service provisioning for customers/end users can be done from one place. Service configurations for the network can also be controlled from the same service platform.

In addition, the concept of separate service profiles will enable the end users/customers to personalize their service in real time or near real time. This will not only enable the customers/ end users to change their preferences through portals or mobile portals, but also enhance the customers'/end-users' experiences. Further details on system functions for service profiling and service control can be found in Section 8.7.4.

8

System Functions and Development

8.1 Introduction

Systems and system (and network) management are integral parts of any service. This chapter details the systems functional areas requiring development when designing a service. The emphasis is on the functions that need to be developed to support a service, not to dictate the systems architecture of the service (i.e. on which systems these functions shall reside). Different companies have different systems (BSSs and OSSs) that perform these functions in their current operational environments. The model in discussion is designed to ensure that the tasks or functions that can be automated are considered when designing the systems to support and operate the service, irrespective of the BSS and OSS that may already be in place. It is up to the system architects within the service providers/operators to decide in which system these functions shall reside. For those interested in architecture designs for OSS systems, I would recommend further reading on this topic in *OSS Guide for Telecom Service Providers and ISPs* [51].

The functions discussed in this chapter are by no means exhaustive as they will vary between services, but should provide you with a good start when designing a service. In the process of going through these functions, I hope it will help you to develop a better set of requirements and trigger thoughts on functionalities that you might need as well as helping you to develop a solution that suits you. The functions listed are at a relatively high-level, but should enable you to tailor them when specifying detailed system requirements. In many cases, you will probably be buying a system off-the-shelf and tailoring the functions to your service. In that scenario, the functions listed in this chapter should facilitate a more complete set of requirements and provide you with a check list of functions to measure your solution against. Some of the functions may not be relevant to the service you are designing, so please feel free to ignore them.

The system functions and tasks detailed in this section can sometimes be performed by operational personnel. The service designer needs to decide which solution is feasible and most cost effective and efficient. Hence, operational processes are to be developed closely in parallel with system functions.

For those who are familiar with New Generation Operational Support System, you may find some of the functions listed in this chapter are detailed in the *Telecom Application Map*

Successful Service Design for Telecommunications   Sauming Pang
© 2009 John Wiley & Sons, Ltd

(TAM) [54]. However, the emphasis for this chapter is the system functions to be considered when designing a service. Nevertheless, I find the *Telecom Application Map (TAM)* [54] to be a useful guide.

8.1.1 Systems Requirements and Methodology

Systems requirements for the service are derived mainly from the service requirements. Requirements for each functional area are covered in each of the sub-sections.

There are many system design methodologies (e.g. Object Orientation and the Structure System Analysis and Design Method) and they are not discussed here. This chapter is meant to cover system functions at a high level for a service, rather than designing the system itself. It is not intended to cover how these functions should be carried out from a system design point of view. It is written in this way mainly because service designers do not necessarily need to know how the systems are designed to provide the required functionalities to support the service. Please refer to software and system design books [21, 30] for system design activities and methodologies when designing the systems to carry out these functions.

8.2 Interrelationships Between the Functional Areas in the Systems Domain

As stated in Chapter 6, the functional areas in the systems domain include:

- customer creation and management;
- order management;
- network provisioning;
- service provisioning;
- call/session control;
- billing, charging and rating;
- service accounting, revenue accounting and OLO bill reconciliation;
- network and service management;
- fault management;
- performance management;
- capacity management, traffic management and network planning;
- system support and management;
- reporting.

The interrelationships between these functional areas to support a service are illustrated in Figure 8.1.

The external factors initiating system functional events for a service normally originate from the customers/end users or operational teams, who operate and manage various operational areas (i.e. from people). All external influences for each functional area are detailed in each of the respective sections.

Within the business support system domain, I have included functions like customer creation and order management, billing, service and revenue accounting and bill reconciliation as business support functions, although some argue that the reporting and fault management functions should be included in the BSS domain. In real terms, both functions are in the operational support domain, as fault management is mainly related to network operations and most of the reports produced (apart from customer reports) are used to improve operational efficiency.

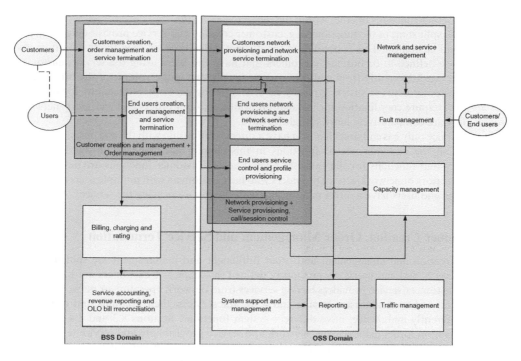

Figure 8.1 Interrelationships between functional areas.

For the operational support system domain, I have separated it into the following functional areas:

- network provisioning;
- service provisioning;
- call/session control;
- network and service management;
- fault management;
- capacity management;
- traffic management;
- systems support and management;
- reporting.

All of this together should cover all the necessary OSS functions. Workforce management should also be part of the operational support system domain. However, this is not service specific and is outside the scope of this book.

The main interfaces between the BSS and OSS domains are the:

- order management and provisioning functions;
- service provisioning and billing functions;
- billing and reporting functions;
- network provisioning and bill reconciliation.

Most of the other functions operate within the respective domains.

I have also split some of the functions (e.g. customer creation and service provisioning) between end users and customers. This is because of the distinct differences between the two entities. As customers are defined to be business customers that have end users using the service, customer orders are treated differently from end-user orders, mainly due to the volume of the orders. For example, customer network orders typically involve ordering network connections for multiple sites and will require coordination for installation, whilst end-user orders and end-users' network provisioning are comparatively more straightforward.

If your service only has end users (i.e. you have a direct relationship with the end users), then you can ignore all the customer-related functions in this chapter and concentrate on developing requirements and functions for the end-user-related tasks. Customer functions are designed for network operators or network/service providers who have customers with multiple users or who do not 'own' the end users/have a direct relationship with the end users using the service.

8.3 Customer Creation, Order Management and Service Termination

Customer creation and management and order fulfillment and management are split into two levels: customers and end users. As previously defined in Chapter 1, customers are business or wholesale customers/service providers buying services from network operators or network/service providers and end users are users buying and using the services; their accounts and orders are handled differently and, hence, have different system functions. Sections 8.3 and 8.4 cover the customer creation, order management, network and service provisioning and service termination functions, while Sections 8.5–8.7 deal with end user creation, order management, service provisioning and service termination functions. Therefore, Section 8.3 and 8.4 are written from the view of providing services for business customers, wholesalers and service providers. This section should be read in conjunction with Section 9.2 in order to gain the full picture for the process and system functions to carry out customer order creation and fulfillment tasks.

8.3.1 Customer Creation and Management

The first step to customer management will be having customer details and creating customers on the system. This information is normally held in the customer management system (commonly known as the CRM system). In some organizations, customer information for both existing customers and potential customers is held in the same system. This helps the sales team to manage sales leads.

8.3.1.1 Requirements

Therefore, when capturing the requirements for the customer management system, the questions we need to ask include:

- Where will the customer information come from?
- What is the information going to be used for?
- How much detail should be held in the system?
- Who will be authorized to use the customer information?
- In what format (e.g. electronic or input from a user of the system) will the customer information be?
- How many levels of authorization will the system need? What are the different levels of authorization for accessing the information (e.g. read only, read/write access, administration)?

- If information is entered manually, which field can be modified by whom (i.e. which level of system user authorization)?
- Does the customer information need to be passed to another system? If so, which system and what details?
- What is the access mechanism for the system?
- Is the system for internal use only or will the customers be allowed to use the system?
- If customer interfaces are required, what information will the customer be able to see and what functions can they use. What security measures are required?
- How long does the system need to hold the customers' information for if the customers are no longer active?
- What and how many statuses/states can a customer have (e.g. potential, active, deactivated)?
- How many customers are expected to be created per day/per week/per month?
- What is the frequency of customer data input?
- Is the system expected to go and get the data from elsewhere?
- Is there a requirement to track sales leads and sales progress for the system as part of the CRM?
- What other customer management functions are expected (apart from, for example, tracking sales leads, knowing a customer's account history, SLA levels, fault history, etc.)?

8.3.1.2 System Functions

From a systems point of view, creating customers means that the system needs to provide the facilities to enable users of the system to create customers on the system. Customer information can come from the sales department; hence, a human interface will be required. Automated customer creation functionality may also be required if the information input is from another system, either with the customer or wholesaler. In this scenario, the file format for the customer information must be specified so that the information can be processed.

To protect the company from bad debts, it is advisable to perform credit checks on all the customers. This information should be held in the system. The credit check request function can be automated; however, some human intervention may be required.

When creating a customer on the system, the following functions should be considered:

- obtain customer details, including billing address, phone numbers and contact person for various functions (e.g. payment, faults/planned work, commercial issues), existing or potential or deactivated customers;
- record customer profile (e.g. market segment, potential revenue);
- select the services being provided or services that can potentially be provided for customers;
- generate account numbers, if sale orders have been generated;
- verify all necessary details are complete;
- check for duplicate accounts;
- ability to modify details where necessary;
- names of person in the organization responsible for the account (e.g. account manager);
- deactivate customer (including reasons) as requested;
- support multiple access mechanisms (e.g. via Web portal or internal network);
- maintain account history (including services, payment history, billing amount, problems);
- management of account (including tracking of current open issues, SLAs and status of potential problem resolution, etc.);
- interface with external credit check system or organization;
- record credit rating/credit limit;

- document management of customer contracts and customer network/service design documents;
- tracking of sales leads and sales progress.

There may be a requirement to data-fill various downstream systems automatically (e.g. billing and reporting systems) with the same information. One needs to decide which system holds the master data. Holding duplicate data on different systems is not a desirable situation, but is sometimes unavoidable. In general, the customer information in all the customer-facing systems should be consistent. One way to achieve that is by populating them all from the same source of information.

If a separate billing system is used, then the following additional information and functions will need to be created:

- Enter customer details as above. The account number in the billing system may or may not be the same as the one in the customer management system.
- Verify all necessary details are complete, including address, phone numbers and contact person for payment, billing cycle start date (if applicable).
- Ability to modify details where necessary.
- Deactivate customer account and produce bill.

Depending on the service, customer reporting may be part of the service. If a separate customer-facing reporting system is used, then the following functions should be considered:

- specify account details (e.g. company name, name and title of person responsible for billing, billing address and account number);
- generate billing account number (if this is not automated);
- configure method and medium of delivery of reports;
- configure reporting intervals of various reports if choices are given.

It may also be a good idea to create the customer on the fault management/customer fault reporting system if such a system exists for the service. This will enable consistent customer information across the customer-facing systems.

As part of the system management function, all exceptions and all completed and failed tasks or functions should be logged for audit purposes. It is suggested that these logs are kept for only a limited period (e.g. 3 months), otherwise this will occupy memory space unnecessarily.

Internal management reports for customer creation may include the following (this is very dependent on the service being designed):

- number of new customers created per month;
- number of deactivated customers per month;
- number of rejects/exceptions;
- repeated exceptions.

Figure 8.2 provides a summary of the customer creation functions to be considered.

8.3.2 Customer Network Design

There are two parts to a service for a customer: the customer network connectivity and the services to be activated on the network. For most business/service provider services, almost all customers have specific network requirements; for example, site locations, IP addressing schemes and so on. Therefore, customer network designs are created to ensure all the customer locations

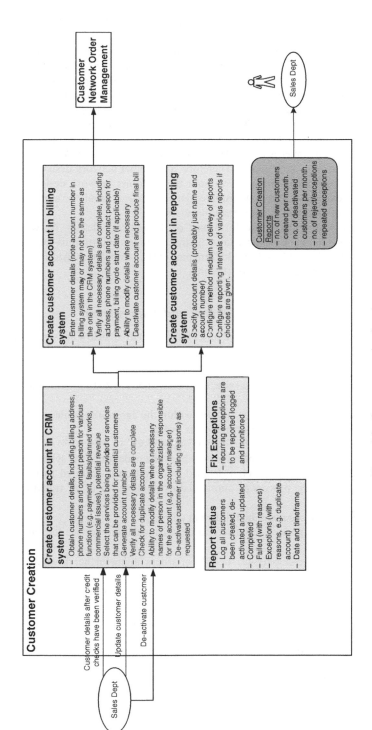

Figure 8.2 Customer creation functions.

for the service are connected. The customer network design forms the basis of where services are being provided.

Network design requirements are heavily dependent on the service in question. The customer network requirements should be set within the framework of the service (as stated in the service description) for the features to be supported. For most fixed-line operators that service the business/corporate sector, the design of the customer network will be done by network designers or a specialist sales team. This is especially true for the IP VPN services, wholesale broadband service, contact center service, etc., as they are complex network solutions. For mobile services, customer network design is less relevant, as the network connections between customers/end users are dynamic, unless the customer is a mobile virtual network operator (MVNO), where various network handover points may be required. Therefore, a customer network design is not normally an automated task for a system. However, without the customer network design, the customer network cannot be provisioned, and that customer network design should be stored in the system for service management and fault management purposes.

A network feasibility study may also form part of the network design. Network feasibility includes core and access network capacity investigation. Choices and options for access technologies and costings will be the result and should form part of the costing in a customer bid. Please see Chapter 9 for the operational processes for this area.

8.3.3 Customer Network Order Fulfillment and Management

When thinking through the customer network order fulfillment area, below are some questions that one should be asking:

- When the customer has ordered the service, what do we need to do to provide the service to the customer?
- How do we ensure the order has been fulfilled within the agreed timescale (i.e. SLA)?
- How do we monitor the status of each customer order?
- What meaningful order status can there be for the service?
- How do we get the order and the information about the order onto the system?
- What should the order include? What detailed information is required to make the order meaningful and make the process as automated as possible for order tracking and service provisioning?
- How do we know if the order can be fulfilled at all?
- What are the different types of order (e.g. orders, cancelation, service termination, change requests)?
- What can the system do to help reduce the risks of orders failing further down the provisioning process?
- What are the items to be checked or validated to minimize customer network provisioning failures and delays?
- Do we always assume network capacity exists? If not, what system functions are required to provide the necessary capacity?
- Do we need to reserve network capacity for the orders before provisioning the customer on the network?
- How many customer orders are expected per week and per month?
- Which operational area is going to be managing these orders and dealing with the order exceptions?
- Is adding new customer sites to an existing network dealt with the same way as new network orders?

- Is moving customer sites dealt with as a new network order for the new site location and a network termination order for the previous site location (i.e. treat as a network reprovide and termination)?

Most of the tasks and functions within the customer network order creation are manual. However, the systems functions listed can assist or prompt the system users to complete their tasks.

8.3.3.1 Order Validation

In general, there can be several types of order:

- ordering new service or service connections;
- modification orders/change requests;
- canceling the service;
- terminating/ceasing existing service.

Depending on the service and the terms of the service, there can be distinct differences between terminating/ceasing and canceling the service. Generally, terminating/ceasing the service means that the service has already been provisioned and the customers do not want to use it any more. Canceling the service can mean that the customers have ordered the service and have changed their minds and want to cancel the order. Normally, there is a time window between the orders being placed and orders being fulfilled. One needs to work out what is the acceptable time window for both the customers and the operator. Once the service has been provisioned, the customers will probably have entered the minimum term of the contract and the billing system will have started billing the customers. Then, canceling the service may be too late. This can get complicated. In some countries (e.g. in the United Kingdom), there is a 'cooldown' period where the customers are allowed, by law, to change their minds and cancel the service within that period. Therefore, cancelation scenarios need to be thought through very carefully. Customer network and service termination are dealt with in Sections 8.4.2 and 8.4.5 respectively.

As part of order validation, the following functions should be considered. Validation at the start of the process is important to avoid order failures due to lack of or invalid information.

- Verification of contract details.
  - o This can be in the form of a contract number or a system user verifying the contract exists and checking a 'tick box' in the system.
- Ensure network design has been signed off by both the customer and appropriate internal design authority.
  - o This is probably going to be a 'tick box' function unless the customer network design is in the form of a template, where the system ensures all the fields are filled in with information of a certain format and range.
- Ensure the customer has been set up in all the customer-facing systems (e.g. billing, reporting and fault management).
  - o This function should be automated where possible to avoid human error and inconsistency.
- Confirm customer credit checks have been performed and passed.
- Verify details on the network order form.
  - o This can be in the form of field validation and ensuring all the mandatory fields have been completed.
- Extract and populate order details/data into the system.

Normally, customer network orders are verified manually, as the customer contract and customer network design document are separate and take different forms. If the network order forms are electronic, then the input format is to be such that data can be extracted and populated into the system automatically. There may also be a requirement to log/attach the customer contract and the customer service and network design documents within the system for future reference.

8.3.3.2 Define Customer Network Order Requests

For each network order, network components are defined and separated so that different network elements and end points can be provisioned on the network. The functions required can be as follows:

- specify access/core network resources per network connections ordered;
- hold order whilst awaiting for capacity if necessary;
- update order status.

The customer's network order might be split into different network sub-orders for all the different network connectivity to be provisioned. As part of this specification, having the updated network inventory will be useful to define the different network connections and check network capacity availability.

Depending on the capacity planning policy for the service, it might be useful to check network capacity availability during the network order connection specification stage. If the network capacity for the service has a 'just in time' availability policy, then the required network capacity for each network connection must be checked (see Section 8.3.3.3 for function details) before the order can proceed any further. If the required network capacity is not available, then the order is put on hold until network capacity becomes available.

If the network capacity policy is assumed to be always available, then there is no need to check network capacity. Once the network connections have been specified and network capacity confirmed, the order can proceed to the customer network provisioning functions.

8.3.3.3 Check Network Capacity

Before network capacity can be provisioned, we need to ensure that it is available, if the network capacity for the service has a 'just-in-time' availability policy. The system functions to support this task may include:

- identify access network locations;
- identify if there is access and port capacity;
- soft allocate or reserve capacity on network inventory (if required);
- check core/end-to-end network capacity;
- check duplicate network capacity requests;
- check OLO capacity (if required).

If the customer network requires network resources from OLOs, then the capacity checks may be beyond the functions of an internal system. Therefore, a manual process may be required. It is not common to have this function totally automated, as the system interface to perform this check with other operators is probably too commercially sensitive. Any capacity shortfall discovered should be fed into the capacity management part of the system.

8.3.3.4 Order Status Checks

Checking order status and maintaining status updates are the essence of order management. Status updates for each order should be automated after the defined checkpoint tasks have been performed and each update should be time stamped. An example network order status includes:

- in progress (with OLO or internal – customer network order specification or internal capacity check or await internal network provisioning, etc.);
- completed (one may choose to inform customers with this order status from the system);
- failed (with reasons);
- exceptions (with reasons).

8.3.3.5 Jeopardy Management

As part of the order management process, jeopardy management plays an important part. Functions within jeopardy management include:

- Identifying and logging failed orders and order exceptions.
- Defining order failure criteria/failed reasons.
- Alertng system users of failed orders.
- Checking the status process of each order.

  o The system should have a defined timeframe for each order status. When the defined timeframe has been exceeded, the order will be in jeopardy and will be at risk of breaching the SLA with the customers.

- Prompting users that the order may exceed the agreed SLA.

The system should have the facility to allow system users to modify orders that failed and to fix order exceptions. When designing the system/service, one needs to decide whether the status of the order can be changed manually. This depends on the amount of control and flexibility required.

8.3.3.6 Network Order Management Reports

A network order management report will be useful for internal monitoring of order progression and help to identify 'bottlenecks' within the operational process. The following reports can be considered:

- number of orders accepted per month;
- number of orders completed per month;
- number of order exceptions per month;
- number of reject/exceptions per month;
- number of outstanding exceptions (with predefined reasons);
- age of jeopardy and 'hold' orders;
- time taken to fix each order exception;
- number of recurring exceptions;
- number of rejected orders (with reasons);
- number of orders on hold due to internal capacity shortfall;
- number of orders on hold due to OLO capacity shortfall.

8.3.4 Customer Network Order Management Summary

The functions for customer network order management are summarized in Figure 8.3.

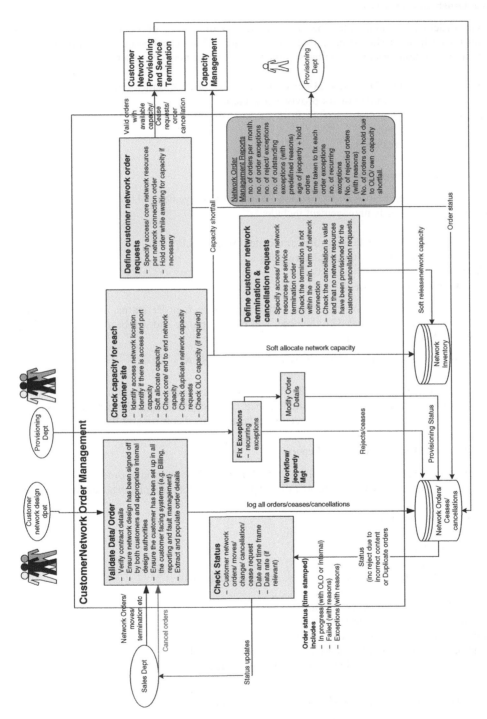

Figure 8.3 Customer network order management.

8.4 Customer Network Provisioning and Network Termination

From a customer network provisioning point of view, one needs to think about:

- What is the maximum numbers of sites a customer might have?
- What is the maximum number of end users a wholesaler/service provider customer is going to have?
- What network elements require provisioning to enable the service?
- What information is required to enable automatic network provisioning?
- What system interfaces are required for provisioning the customer network?
- Can the provisioning be batched up?
- Is there any limitation as to how many customers can be provisioned/deprovisioned on a certain piece of network equipment?
- What is the projected call/session volume?
- Are there any requirements to activate any call holding facilities?
- What are the network elements requiring provisioning in order for the customers to obtain the service?
- How much validation of available network resources/capacity is required before network provisioning activities should take place for each customer?
- How will the system activate network connections? What mechanisms are available to the system from the network point of view?
- What are the protocols to be used by the systems to communicate with the network elements from a network provisioning/deprovisioning point of view?
- What are the network testing procedures to be carried out to ensure all network provisioning activities for the service are completed successfully?

8.4.1 Customer Network Provisioning

Following the definition of the network connectivities for the customer network (as described in Section 8.3.3.2), the network provisioning functions can take place.

8.4.1.1 Validate Network Order Requests

Before network provisioning can start, we need to verify that all required network data is complete and that all parts of the service have been specified. This can be in terms of ensuring that all the template information defined has been completed and the systems checks for all values in all the different fields, parameters and attributes are valid. Once the order requests have been verified, the status of the order should be updated.

8.4.1.2 OLO Network Provisioning (If Required)

It is not uncommon to have parts of the network (especially for the access network) to be provided by another operator. If this is the case, then the system function will need to include preparation of network orders for OLO network connectivity.

Before sending an order to an OLO, it will be wise to have system users of the right authority to authorize these orders before placing them with the OLO, as the financial exposure of these orders might be quite high. Having the system place these orders without necessary authorization may potentially result in unnecessary expenditure.

The above function assumes that there is a systems interface for placing orders and obtaining order statuses from OLOs. If the systems interface does not exist, then some of the task above will need to be performed manually.

8.4.1.3 Process Internal Network Provisioning

If OLO network requires provisioning, it is desirable to obtain confirmation that the OLO network has been activated before proceeding to provision the internal network as this can have a longer lead-time. This is assuming that the internal network provisioning can take place shortly after the OLO network has been activated. Otherwise, you might be paying for OLO network unnecessarily. System functions to be considered for provisioning the network includes:

- check customer network order requests – proceed with internal network provisioning when a predefined status (e.g. ready for internal network provisioning) has been reached;
- ensure all required OLO orders are completed successful;
- check for cancelation request for each customer site before proceeding to network provisioning;
- check all CPE is in place; this is probably a human task, and a 'tick box' function is required for each site;
- provision customer on access and core network according to network resource allocated;
- check field resources if engineer visits are required;
- configure CPE;
- update network inventory;
- update network management system with network provisioning details for customer network equipment that is being managed;
- test end-to-end network connectivity for all customer sites;
- raise network provisioning exceptions if any of the above tasks are not completed successfully.

The configuration of CPE is normally prepared by the operational team based on a predefined template. It may be desirable for the system to generate this configuration. This is dependent on the service and if it is feasible as a system function.

Testing the end-to-end network connectivities may involve having an engineer on customer sites. Therefore, the availability of field engineers is to be checked before the customer installation dates can be confirmed. This is, of course, dependent on the service and the level of configuration required on CPE.

8.4.1.4 Process OLO and Internal Rejects/Exceptions

As part of any automated process, exceptions and rejects will require human attention. During the system design process, a list of possible exceptions should be stated and an operational process will be required to deal with these exceptions from the system. Below are some system functions that will help operational personnel with order rejects and exceptions:

- define reject reasons and reject criteria for each reason;
- identify order rejects and group together in accordance with reject reasons;
- update orders with predefined reject reasons;
- automatically fix reject orders within system where possible;
- generate report for order requiring manual processing;
- GUI for system users to edit and deal with order exceptions (both internal and OLO orders).

8.4.1.5 Customer Network Provisioning Reports

Customer network provisioning reports will be useful to ensure that customers' SLAs are met. The following provisioning reports can be considered:

- number of network provisioning orders completed per week and per month;
- measure the number of orders completed against SLA;
- number of OLO orders completed per week;
- measure the number of OLO orders completed against SLA with OLOs;
- time (average and maximum) taken to fix each order exception;
- number of recurring order rejections and exceptions grouped by rejection/exception reasons.

8.4.2 Customer Network Termination and Cancelation

Network terminations can be for various reasons. For example, the business customer decides to close the premise or one of their offices. Sometimes, depending on the service definition, moving customer locations can be dealt with as a termination and a new order (sometimes known as 'cease and reprovide'). Systems functions to support network termination are not dissimilar to that of network deprovisioning (reverse of provisioning); however, there are some subtle differences. Please note that terminating the network connection does not equate to terminating the service. Please see Section 8.5.5 for details on service termination.

When designing the customer network termination and cancelation functions, one needs to think about:

- Where will the termination/cancelation requests come from?
- Who can raise a termination/cancelation request?
- How will the system know if the network connection is within its minimum term of contract?
- If the network connection is terminated within the minimum term of the network contract, what charges are there?
- Will these charges be pro-rated or a fixed amount?

The main difference between termination and cancelation is that cancelation normally refers to network connections that have been ordered but not yet provided. For most services, there is a cancelation period within the contract, where the customers can cancel without incurring any charges. Beyond that period, the customers will be charged either a fixed sum or the term of the contract.

8.4.2.1 Define Customer Network Termination/Cancelation Requests

For each customer network termination/cancelation request, we need to define what part of the network requires termination and if there is any reconfiguration work required on the rest of the customer's network. To support this task, the following function may be required:

- Specify access/core network resources per network termination and cancelation order;
- Check that the termination is not within the minimum term of network connection.

  o otherwise, the system users need to be alerted.

- Check that the cancelation is valid and that no network resources have been provisioned for the customer cancelation requests.

8.4.2.2 OLO Network Termination and Cancelation (If Required)

This is only required if the network termination/cancelation is provided by the OLO. Functions to be considered are:

- create termination and cancelation requests to be sent to the OLO, including the date of termination;
- verify OLO termination/cancelation requests created;
- create termination and cancelation file to be sent;
- obtain authorization from operational personnel for all requests before sending orders to the OLO;
- send termination/cancelation requests to the OLO to the predefined destination;
- log termination and cancelation requests that are sent to the OLO;
- obtain status update from the OLO with respect to the requests;
- update status of termination and cancelation requests.

8.4.2.3 Process Internal Network Termination and Cancelation

Functions to support the customer network termination and cancelation may include:

- deprovision customer network connections on access and core network according to network resource specified;
- ensure that the network is deprovisioned on the specified date;
- cancel the relevant network provisioning order;
- reconfigure the rest of the customer network where necessary;
- update network inventory to release network resources;
- update date billing system.

8.4.2.4 Process OLO and Internal Rejects/Exceptions

The system functions to process termination exceptions are the same as that of network provision as stated in Section 8.4.1.4, though the human actions required will be different.

8.4.2.5 Customer Network Termination and Cancelation Reports

The following reports may be useful for the customer:

- number of network termination requests received per month;
- number of network cancelation requests received per month;
- number of network terminations requests completed on specified dates on a per month basis;
- number of network cancelation requests completed per month;
- number of OLO network termination requests completed within the required date;
- number of OLO network cancelations completed and measured against SLA with the OLO;
- time taken to fix each termination exception;
- number of recurring termination exceptions, grouped by exception reasons.

8.4.3 Customer Network Provisioning Summary

A summary of the customer network provisioning functions is provided in Figure 8.4.

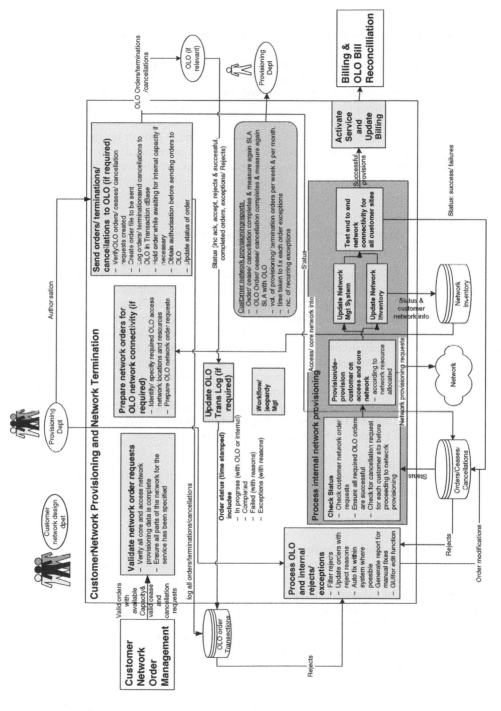

Figure 8.4 Customer network provisioning and network termination.

8.5 Customer Service Provisioning (Including Moving, Additions and Changes)

There are distinct differences between customer service provisioning and customer network provisioning. Customer network provisioning enables the customers to gain network connectivities between their sites; that is, where information/network traffic is passed. Service provisioning enables services/applications to run on the network that has been provisioned and allows customers to perform the functions or transactions as required. The service layer is above that of the network layer, where certain service provisioning activities may then be translated to numerous network profiles/attributes for activation. For example, enabling a VoIP service between customer sites will involve activating the correct QoS profile in the network in order to provide the right treatment for the voice traffic. Therefore, service activation is very much intertwined with that of network provisioning and profile activation. However, both activities are distinctly different, and the basic network connectivity provisioning will need to take place before service provisioning activities can occur.

8.5.1 Customer Service Profile Control (Especially in QoS-Based Services) and Service Activation

Customer service control functions allow different service profiles (i.e. different network parameters) to be set to create different types and grades of service. For example, if you were running a voice service over a PS data network, then you would want to be able to prioritize the voice traffic over the data traffic. Therefore, a different QoS profile for the customer network is to be set to enable that to happen. Similarly, in the mobile world, if the customer has subscribed for both data and voice services for their end users, then appropriate service profiles are to be applied to ensure that their services are guaranteed. The amount of system functions for customer service control depends on the architecture (centralized or distributed) of the service – that is, if the control of the service is from one place (centralized) or information about service control is stored at the devices at the edge of the network (distributed). The system functions listed below should give you a guide as to what system functions may be required. The functions below should cater for both service architectures. System functions for customer service control may include:

- defining service profiles;
- defining the network parameters (e.g. QoS parameters, security policy) for each of the services/ service packages/service profiles.

System functions for customer service activation may include:

- check network provisioning status;
- obtain the required network parameters from service profile database (if appropriate);
- obtain customer network details (e.g. customer network IP address for data service or customer phone numbers for voice service) from network inventory (if not automated, this should be part of the customer network design and details are to be manually entered onto the system);
- specify network elements to be provisioned;
- provisioning the service-specific network parameters on the appropriate network elements;
- test and confirm the service profile applied is functional;
- update status of service orders;
- activate billing.

For service profile definition and management system functions, please see Section 8.7.4.

8.5.1.1 Security Management (Authentication and Authorization)

Security management for the customer's network is very important to avoid intruders gaining access to a customer's network or data. From a customer service provisioning point of view, the main system functions are to provide the ability to define and apply security policies/method of authentication and authorization on a per customer site basis as part of the service profile to be applied to the network.

8.5.1.2 QoS Services

In today's services/applications, having the ability to define and apply QoS parameters to the service and service network elements is probably one of the most important features required from the customer service provisioning system point of view. Such functionalities are required for services that mix delay-sensitive applications/traffic (e.g. voice, TV and to some extent gaming) with those that have higher tolerance levels on delays (e.g. data transfer application – e-mail, file transfer, Web pages download) in the same underlying packet-switching network infrastructure.

8.5.1.3 Billing Activation

Before billing is activated, we need to ensure that the service testing is successfully completed for all the customer sites. Depending on the service, it may be little use for the customer if the service is only connected at two of their sites (out of 10). In addition, if you start billing the customer for a half-provisioned service, they will not be very happy.

8.5.2 Moves, Adds and Changes

Moves, adds and changes (MACs) are potential service requests from the customers. These should be defined as part of the service definition. Below are some common examples of MACs requests.
Changes requests could include:

- changing service profiles (e.g. upgrading service package to include additional services or downgrading service to lower price plan);
- change of company name due to takeover or buy out;
- change of company's billing address.

Moves include:

- moving network termination equipment to a different location in the same building;
- moving site to a different geographic location.

Adds includes:

- Adding more services (e.g. VPN + IP TV services) to the existing ones. This could be classed as a change to the service profile. The method of application of this addition depends on the service defined.
- Adding more security policies to the existing ones.
- Adding additional sites. This can be classed as a new order, depending on the service definition.

Therefore, when designing the systems functions for service provisioning, one needs to consider all the scenarios above.

8.5.3 Customer Service Provisioning Exceptions/Jeopardy Management

System functions for handling customer service provisioning exceptions include:

- Defining customer service provisioning exceptions and remedial actions to be carried out on each of the exception scenarios (e.g. retry performing the task or alert system users of exceptions).
- Logging customer service provisioning exceptions.
- Checking the status of each customer service provisioning order.
- The system should have a defined timeframe for each order status. When the defined timeframe has been exceeded, the order will be in jeopardy and will be at risk of breaching the SLA with the customers.
- Prompting users the order may exceed the agreed SLA.

The system should have the facility to allow system users to modify customer service provisioning orders that failed and fix order exceptions so that the order can progress with the required customer provisioning activities.

Scenarios for where exceptions might be generated from the customer service provisioning tasks may include:

- unable to obtain the required service profiles;
- unable to obtain the required network parameters for the service profile;
- unable to apply network parameters on certain network elements;
- invalid customer service cancelations;
- invalid customer service change requests;
- orders about to exceed agreed SLA.

8.5.4 Customer Service Provisioning Reports

The following customer service provisioning reports may be of use:

- number of service provisioning requests completed within SLA per month;
- number of service change requests completed by the requested date per month;
- number of service provisioning exceptions;
- average time taken to fix each service provisioning exception;
- number of recurring service provisioning exceptions.

8.5.5 Customer Service Termination and Cancelation

Service termination is different from network termination. The customer may still have a service, but part of the network may be terminated due to moving of site locations and so on. However, if the customer terminates the service, the network components related to the service will need to be terminated at the same time.

Cancelation can be dealt with as a service termination, depending on the service. One needs to be aware that if customers cancel one service, then this does not mean that they are canceling the other services they may have. Depending on the service, the customer may be able to cancel the service within a period of time after placing the order. Cancelation charges may apply. Termination of the service is normally subject to a minimum term of the commercial contract. When designing the system, we need to decide how these charges can be checked and billed.

System functions related to service termination and cancelation includes:

- Check if the customer is eligible for cancelation. Otherwise exception should be raised.
- Check the termination is not within the minimum term of service (and network connections (if applicable).
- Charges will normally be incurred if the service or the network was terminated within that period.
- Specify all the network resources per service termination/cancelation order.
- Specify date of termination/cancelation.
- Obtain network parameters for service to be terminated/canceled from service profile database. If the customer is canceling one service and keeping another service, then the service profile of the remaining service is to be obtained.
- Obtain customer network details from network inventory.
- Specify all network elements to be deactivate for service termination.
- Specify all network elements to be changed for service cancelation.
- Reprovision the network parameters in the appropriate network elements for the remaining service (if required).
- Deactivate network parameters in the appropriate network elements for the service to be terminated on the specified date.
- Initiate customer network termination functions (if required).
- Update order status.
- Update billing regarding the completion of service termination.

If the customer is also terminating the network contract, then the functions within the customer network termination can be used to perform the network termination tasks after the service has been deactivated on the network elements.

8.5.5.1 Customer Service Termination and Cancelation Reports

The following customer service termination reports may be of use:

- number of service terminations completed by the request date per month (groups with predefined reasons);
- number of service cancelations completed per month (groups with predefined reasons);
- average time taken to fix each service termination exception;
- average time taken to fix each service cancelation exception;
- number of recurring service termination exceptions;
- number of recurring service cancelation exceptions.

8.5.6 Customer Service Provisioning Summary

A summary of the customer network provisioning functions is provided in Figure 8.5.

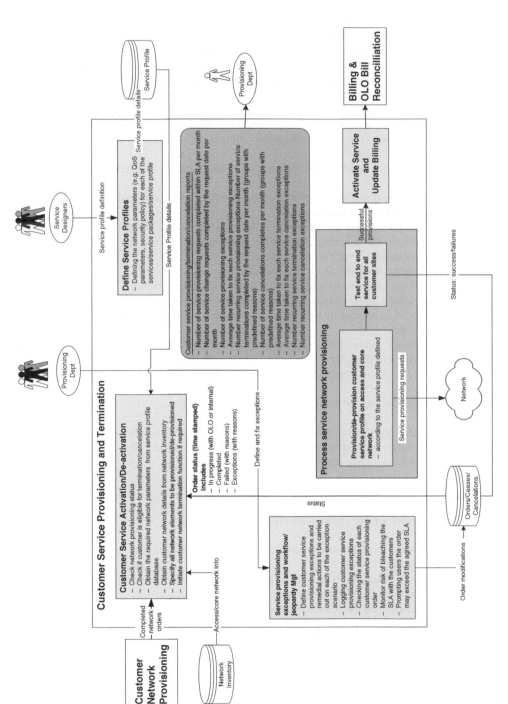

Figure 8.5 Customer service provisioning summary.

8.6 End-User Creation and Order Management

When designing the system functions for end-user creation and order management, you might want to ask yourself the following questions:

- Do we need to know the end-user details for the service? If so, what details do we need and how much detail do we need to store?
- What level of detail can be stored before getting into any privacy issues?
- Where is the end-user detail going to come from?
- Where will the end-user orders be coming from? What are sales channels for generating end-user orders?
- What tools are required for end users to sign up for the service by themselves? If it is a Web-based tool, what security measures are required to ensure authenticity of the end users and the security of the system/tool provided?
- How many new users are expected to sign up for the service on a daily/weekly/monthly basis?
- How often will end-user orders arrive?
- What types of end-user orders are there?
- Are the end users allowed to cancel the service within a period of time after signing up for the service?
- Are these end-user orders processed in batches or as they arrive?
- Do the end users expect acknowledgements and updates for their orders?
- Will the end users be able to 'view' their order progress?
- Do the customers expect acknowledgements and updates for their end-user orders?
- What is the SLA for provisioning the end-user orders?

There can be several scenarios for end-user orders, depending on the service on offer and in which part of the value chain the service is:

1. End users signing up for the service with the operator direct.
2. End users signing up for the service with the service provider and service provider is buying part of the packaged service from a network provider (e.g. end user signing up for a broadband service with an ISP where the operator is providing the access network connectivity or mobile end user signing up with an MVNO for mobile services or end user signing up for a mobile phone service through resellers).
3. End users of a business customer using the service provided by the operator to the business customer (e.g. the employee of the company using the mobile service where the company bought the service from the operator).

The following sections are written from the point of view of an operator providing a service for service providers (i.e. scenario 2 above). This scenario is used as it is probably the most complicated one of all. Therefore, when reading through the following sections, you should adapt these principal functions to the service you are designing and use the following sections to stimulate thoughts on system functions relating to end users and their orders for the service. For a mobile operator, some of the functions may not apply and they will be highlighted in the relevant sections.

8.6.1 End Users and Orders Creation

End-user orders can be initiated by the customers or end users themselves, depending on whether the service is in the business/wholesale environment or retail environment (as listed in the section above).

End users may want to order the service over the Internet, in which case a Web-based tool will be required and end-user authentication and security of the system will need to be considered. If the service is sold through resellers, then system interfaces for the resellers will be required.

The creation of the end users on the systems depends on whether you directly 'own' the end users and if you bill them for the service directly or on whether the end users are acquired through a third party.

End-user creation functions may include:

- create end-user account (if appropriate);
- verify that all necessary end-user details (e.g. name, address, postcode) are complete;
- record end-users' demographic details;
- support a multiple access mechanism (e.g. via Web portal or internal network);
- generate phone number, user IDs or USIM/mobile station international subscriber directory number (MSISDN) for mobile service;
- management of account history (including services, payment history, billing amount, problems);
- interface with external credit check system or organization;
- record credit rating/credit limit.

If you directly 'own' the relationship with the end user, it is recommended that credit checks are carried out on the end users before order acceptance.

8.6.1.1 Validate Data/Order

Whichever way the end-user orders are arriving, they need to be validated. The following functions are to be considered:

- validate data integrity with acknowledgements;
- check for duplicated orders/moves/change requests;
- simple field/order information validation (name, caller line identity (CLI; if any), address, post code, etc.);
- if orders are in batches, separate batch orders into individual end-user orders/cease/cancelation requests with acknowledgement to each end-user order;
- validate whether end-user requests already exist.

One also needs to think about what order types are allowed over which channel for the service. For example, the end user can only cancel or terminate the service by ringing the help desk because the service provider may be keen to keep their end users by offering them a different deal when they speak to their help desk.

For a fixed-line service (e.g. PSTN and broadband services), one needs to consider end users moving house. With some operators, moving house is dealt with as 'termination' and 'reorder'; this might be fine in principle, but what happens if they move within their minimum term of service? Service termination and cancelation are discussed in Section 8.6.2.

A change request is another order type to be considered. Change requests could be changing names of end users, changing addresses (not for fixed-line services, as this might involve a move request, as discussed above) or changing (upgrading or downgrading) service profiles. All the change-request scenarios are to be thought through when designing the service and the system functions. Service change requests are covered in Section 8.8.7.

8.6.2 Define End-User Orders/Change Requests

For each end-user order, network and service components are defined and separated so that different end points can be provisioned on the network and the service profiles of the end users are set correctly. Therefore, the functions required can be as follows:

- specify access/core network resources per end-user network connection order;
- specify service profiles per end-user order;
- specify the change for change requests;
- hold order while awaiting capacity if necessary;
- update order status.

Capacity or network resource checks probably only apply to the fixed operator environment. For mobile operators, checking the availability of a handset requested by the end users is more relevant here.

8.6.3 Check Network Capacity

This is most relevant for fixed-line services. We need to confirm that the end-users' addresses are covered by the access network or other operators' networks. This is particularly relevant for a wholesale broadband service, where the end-users' addresses may have broadband network coverage by the operator's own network or by other operators' networks that are part of the service. If the capacity or network coverage is not available, then the order should be put on hold or canceled. The system functions to support this task may include:

- identify access network location;
- identify whether there is access and port capacity;
- soft allocate or reserve capacity on network inventory (if required);
- check core/end-to-end network capacity;
- check duplicate network capacity requests;
- check other network operators' availabilities (if required).

If the end-users' network requires network resources from another network operator (OLO), then the capacity checks may be beyond the function of an internal system. Other system interfaces may be required. If there is a capacity shortfall, then this information should be fed into the capacity management part of the system.

8.6.4 Order Status Checks

The status of the orders should be updated automatically after various functions have been performed or after the defined check point tasks have taken place. It is essential to update and maintain the status for each end user order. Each update should be dated and time stamped. Example end users order status may include:

- in progress (with OLOs (other network operators) or internal – customer network order specification or internal capacity check or await internal network provisioning, etc.);
- completed (with date and time stamp) and inform the end users;
- failed (with reasons);
- exceptions (with reasons).

8.6.5 Jeopardy Management

End-user orders may fail during the end-user order management process. The system needs to be able to identify these failed orders and highlight them to system users/operational personal. Functions within jeopardy management include:

- Identifying and logging failed orders and order exceptions and stating order failure reasons and exception scenarios.
- Alert system users of failed orders.
- Checking the status process of each order.

    o The system should have a defined timeframe for each order status. When the defined timeframe has been exceeded, the order will be in jeopardy and will be at risk of breaching the SLA with the customers.

- Warning system users that the order may exceed the agreed SLA.

    The system should have the facility to allow system users to modify orders that failed and to fix order exceptions. When designing the system/service, one needs to decide whether the status of the order can be changed manually. This depends on the amount of control and flexibility required.

8.6.6 End-User Order Management Reports

In order to manage the SLA for order provisioning and monitoring of order progression, internal order management reports are essential. These reports are also useful for identifying 'bottlenecks' within the operational process. The following reports can be considered:

- volume of end-user orders per day, per week and per month;
- number of end-user order exceptions;
- number of end-user orders being rejected (with reject reasons);
- number of outstanding end-user exceptions (with predefined reasons);
- age of end-user jeopardy and 'hold' end-user orders;
- time taken to fix each end-user order exception;
- number of recurring end-user exceptions;
- number of rejected end-user orders due to internal capacity shortfall;
- number of end-user orders on hold due to internal capacity shortfall;
- number of end-user orders on hold due to OLO capacity shortfall.

8.6.7 End-User Order Management Summary

A summary of the end user order management functions is provided in Figure 8.6.

8.7 End-User Network Provisioning

End-user network provisioning is defined to be providing the basic end users with network connectivity. Functions to be performed to activate a network profile or attributes as part of the service will be included in Section 8.8 (on end-user service provisioning).

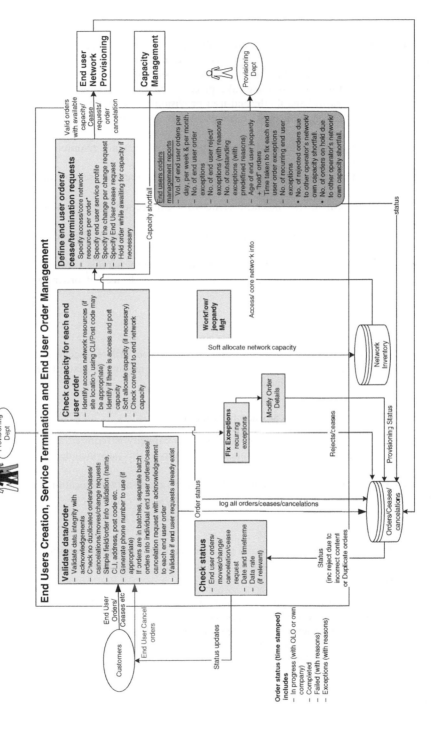

Figure 8.6 End-user creation, service termination and end-user order management summary.

When thinking through the network provisioning activities, one may consider asking the following questions:

- What network elements require provisioning for the end-user's service?
- What system interfaces are required to provision the end users?
- What are the system user interfaces for network provisioning?
- How often can end-user provisioning take place, given that the network elements will be handling real-time traffic? Should the end-user provision be batched up?
- Is there any limitation on the network element on how many end users can be provisioned at any one time?
- When provisioning an end user, what configuration changes are required on each of the network elements?
- What mechanism or protocol will be required to provision the end users on various network elements?
- Is there a requirement to test the end-user network connection after the network has been provisioned?
- Does the end-user network provisioning include the provisioning of the end user's own network devices?
- How do we ensure the end-user's device is compatible with the service network?
- Is there a requirement to remote upgrade the end-users' devices in order to activate the network connection and service?

8.7.1 End-User Network Provisioning

It is important that all the network information for all the end user orders is accurate and complete before passing through to the network provisioning stage. Any missing or inaccurate information will cause provisioning failures and incur unnecessary delays.

8.7.1.1 Send orders to other network provider (if required)

If it has been identified that another network provider's capacity is required, an order for that network resource will be required. System functions to support this may include:

- Identifying/specifying required other network provider's access network locations and resources.
- Creating orders to be sent to other network providers.
- Creating order file to be sent.
- Verifying orders requests created.
    - Depends on the volume of orders and the financial exposure of each end-user order; these orders may need authorization before sending. It may be decided that random checks by operation personnel will suffice.
- Obtaining authorization before placing orders with the other network operator (if required).
- Sending orders to other network providers (OLOs).
- Logging orders and transactions that have been sent to other network providers (OLOs).
- Holding order while awaiting for internal capacity if necessary.
- Obtaining status updates from the other network providers (OLOs) for all orders placed.
- Recording and updating network provisioning information from the other network providers.
- Updating status of end-user orders that have been sent to other network providers.
- Providing the facilities for system users to view and amend orders to be sent to other network operators.

It is assumed that there is a system interface for placing orders and obtaining order status with other network providers. In the absence of a system interface, manual or semi-manual operations might be required.

8.7.1.2 Process Internal Network Provisioning

Where there is no external network involved, the end-user orders can progress to internal network provisioning once the orders are accurately defined. If another network operator's resources require provisioning, then it is desirable to confirm that these have been activated before proceeding to provision the internal network, as that tends to have a longer lead time and information from the external network may be required to complete the internal network provisioning tasks. This is assuming that the internal network provisioning can take place shortly after the OLO network has been activated. Otherwise, you might be paying for an OLO network unnecessarily. System functions to be considered for provisioning the network include:

- check end-user network order requests if they are at a predefined status (e.g. ready for internal network provisioning);
- ensure that all required other network providers' orders are completed successfully;
- check for cancelation request from end users before proceeding to network provisioning;
- provision end user on access and core network according to network resource allocated;
- update network inventory;
- update end-user account with network information (if appropriate);
- test end-to-end network connectivity (if possible);
- raise network provisioning exceptions if any of the above tasks are not completed.

Testing the end-to-end network connectivity may involve having an engineer at the end-user site. This is, of course, dependent on the service and the level of configuration required on end-user devices.

8.7.1.3 Process OLO and Internal Rejects/Exceptions

As part of any automated process, exceptions and rejects will require human attention. As part of the system design, a list of possible exceptions should be stated and an operational process will be required to deal with the exceptions from the system. Below are some system functions that will help with order rejects and exceptions:

- identify order rejects filtered and grouped with reject reasons;
- update orders with reject reasons;
- automatically fix rejects/failures within system where possible;
- generate report for manual fixes;
- GUI for system users allowing them to editing orders (both internal and other network operator orders).

8.7.1.4 End-User Mobile Network Provisioning

End-user network provisioning is defined to be providing the end users with network connectivities. All service-related items, like QoS, session control, etc., are included as part of end-user service provisioning (covered in Section 8.8).

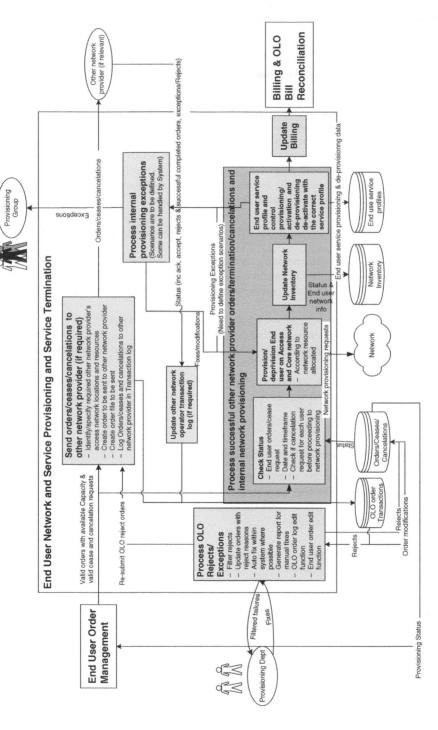

Figure 8.7 End-user network and service provisioning summary.

For a mobile network, the network activation involves registering the USIM/IMSI/MSISDN in the HSS. The provisioning of network resources per end-user session is allocated dynamically as the mobile users go between different geographic locations. Therefore, no access or core network preprovisioning is required for mobile end users.

8.7.2 End-User Network and Service Provisioning Summary

A summary of the end-user network and service provisioning functions is provided in Figure 8.7.

8.8 End-User Service Provisioning, Service Control (Especially in QoS-Based Services) and Service Termination

End-user service profile/service activation provides the control for the end-user sessions. End-user service profile and session policies will be translated to network profile/attribute activation. For example, if the end user only subscribes to a browsing service, then the QoS setting applied for the session will be different to that of the end user who has subscribed to the streaming service.

8.8.1 Service Profile Definition and Management

As services and service bundles/combinations are becoming more diverse, it is essential to manage the services, the service portfolios and service profiles offered to the customers/end users from one place. The service profile defines control policies and the parameters by which the service requirements will be fulfilled, both from the network/system point of view and from an end-to-end service/SLA perspective. These service profiles and control policies define the end-user's service and the control mechanism/parameters for the end-user's sessions/calls. The service portfolio is the collection of service profiles for the service. The hierarchy of service portfolio, service profiles and policies is illustrated in Figure 8.8.

Within the service profiles, different policies relating to the network, application type, usage and session controls, etc. to support the service can be defined. These policies are then

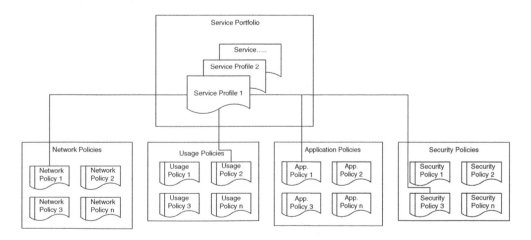

Figure 8.8 Hierarchy of service profiles and policies.

translated into different network/application parameters/settings; for example, network para-meters/settings for treatment of network traffic (QoS policy) or application parameter (e.g. using attribute value (AV) pairs) to provide the service for that policy. This combination of policies and parameter settings bridges the gap between the service that is being offered to, and understood by, the customers/end users and the network and system capabilities and configuration required. Within the service portfolio/profile application and database, the following functions and data should be considered:

- Define different service profiles; for example, for a video service, the QoS, level requirements.
- Provide the ability to view service portfolio; that is, the collection of service profiles on offer, what the current and past service profiles on offer are and their QoS/SLA parameters/policies.
- Define policies from a network perspective and define the application types supported; for example, define the network parameters for each network element for each application server/service for each QoS level on offer. Policy definitions can include:
  - network QoS levels (e.g. network parameters for each QoS level);
  - application policy (e.g. parameters for each application or allowing access to certain content – on a one-off basis or within a certain time window);
  - session control and usage policy (e.g. daytime use only or limited download of 1 Gbit of data per month or the maximum length of each call or session);
  - security policy (e.g. the verification and authentication procedures required);
  - credit control policy (e.g. credit control to limit the length of call or session that each end user cannot exceed);
  - privacy policy (e.g. who is authorized to access end-user's location information);
  - SLA support policy (e.g. fault resolution within 4 h for gold service).
- Translate the different service policies into a form that can be understood and enforced by each of the network elements and systems providing the services.
- Define service inventory; for example, network elements involved to provide the different services.

When defining new network policies, it is important to ensure they can be enforced without jeopardizing the network operation and without introducing a negative impact on current services. These network policies are important, as they set the control policies by which the different network traffic types are treated in the network.

8.8.1.1 3G Mobile Service Profiles

For 3G mobile services, the service portfolio can be very diverse and each service can have multiple relationships between network providers and content providers. These relationships and their subsequent revenue apportionment need to be carefully managed. Hence, it is useful to have all the information in one place to give a clear view of the service portfolio.

In addition, end users might require different services or different preferences to their service when travelling abroad or visiting/roaming to another network. The concept of service profiles will enable these types of end-user personalization. These capabilities also enable end users to change their preferences through mobile portals.

Within the UMTS framework, QoS levels are defined in terms of authentication of USIM/IMSI, admission control, authorization of different services, resource reservation, bearer service levels/classifications, traffic shaping and scheduling and marking of different traffic types and so on. These could be defined as part of the policies listed above.

Additional policies for mobile end users may include:

- Authorization function to inform the access bearer of the set of services allowed for that end user and the gateway that it needs only to forward packets for those services.
- End users' charging options (online prepaid credit checking and offline charging/billing).
  - The online prepaid credit checking involves the checking and monitoring of available credit per end user for the service requested. The system may need to allocate a 'credit quota' for the access bearer during the establishing of the session. Appropriate actions need to be defined for the service if the end users have insufficient credits for the service session (e.g. do you give the end user an 'overdrawn' limit, or do you inform the end user of the credit situation and terminate the session?)
  - Offline charging will involve the collection of charging data and billing the customer according. Please see Section 8.10 for these functions.

Further information on service portfolio/architecture and end-user profile data definition can be found in *3G Multimedia Network Services, Accounting, and User Profiles* [52].

8.8.2 Security Management (Authentication and Authorization)

Security management of end users and end-users' sessions is important to avoiding the misuse of identity or service, obtaining of information illegally and fraud prevention. The most common form of security management is authentication of end users and authorization of the use of service and information to be obtained. This could be defined as part of the security policies within the service profile. System functions for end-user security management include:

- define end-user authentication mechanism/protocol used;
- define where end-user authentication requests are to be sent to for validation;
- define authorization mechanism/protocol;   .
- define the different authorization levels for end users.

8.8.3 Privacy Management

Privacy control system functions can be performed as part of the security management authorization functions above. This enables each end user to access only information/data that is appropriate to them. This is especially important for information relating to customers/end-users' personal details, as unauthorized access to this data will open doors to fraudulent activities (e.g. identity theft). One other application for privacy management policy is the ability to apply parental control over which websites their children are allowed to access.

For mobile services, one should be aware that location information of customers/end users should be kept confidential, as customers/end users probably will not want to disclose their whereabouts or like the idea that their whereabouts are being tracked due to privacy issues. This is an especially important issue for location-based services. The privacy management/policy should protect this type of information from being obtained by unauthorized parties.

8.8.4 End-User Activated Services/Service Profiles

System functions for end-user services/service profiles activation may include:

- check end-user network provisioning status;
- obtain end-user network/device details (e.g. the network element that the end user's profile is to be applied to or end-users' phone numbers);

- obtain the required network parameters (e.g. QoS parameters, site/call/session authorization mechanism – please see the sections above for further examples) from service profile database;
- activate the service-specific network parameters on the appropriate network elements;
- update status of service orders.

Depending on the service, the service profiles may be dynamically applied to each of the end-user sessions, as the service may allow end users to subscribe to certain services on a one-off basis or within a certain time period.

The provisioning of a mobile end-user's service will involve the activation of USIM/IMSI in the HSS/IMS. Please see *The 3G IP Multimedia Subsystem (IMS): Merging the Internet and the Cellular Worlds* [45] for details.

8.8.5 End-User Cancelation

One needs to decide when an end user can cancel a service during the provisioning cycle. In general, it will be neat for users to cancel the service within a set period of time after placing the order. It will be undesirable to cancel the service when network and service provisioning activities have been completed and the end users are billed for a canceled service. However, in some countries (e.g. in the United Kingdom), there is a 'cooldown' period where the end users have the right, by law, to cancel the service within that period. Therefore, cancelation scenarios need to be thought through very carefully. The system function to support end-user cancelation is the same as the end-user termination. Please see Section 8.8.8 for details.

8.8.6 End-User MACs

MACs for the end users should be defined as part of the service definition. Below are some example scenarios one can consider.

Change could include:

- changing service profiles (e.g. upgrading service package to include additional services or downgrading service to lower price plan);
- change of name due to change in personal circumstances (e.g. getting married or divorce);
- change of billing address – this excludes moving house (please see the move items below).

Moves may include:

- moving house;
- moving end-user equipment to a different part of the house.

Adds may includes:

- Adding new services (e.g. Web browsing + IP TV services) to the existing one. This could be classed as a change to the service profile. The method of application of this addition depends on the service defined.
- Adding parental control to end-user accounts.

8.8.7 End-User Network Service Termination

End-user service termination will involve terminating the network connectivity as well as deactivating the end-user's service profile.

When thinking though end-user network service termination functions, you might want to consider the following questions:

- What channels will the end-user's service termination notification come from? Same channel as orders?
- Under what circumstances are the end users allowed to terminate the service?
- How many end-user service terminations are expected on a daily, weekly monthly basis?
- How often do end-user termination notifications arrive?
- Are these end-user terminations processed as batches or as they arrive?
- Will the end user be able to pick a date to terminate the service?
- Do the end users expect acknowledgements and updates for their terminations?
- Do the customers expect acknowledgements and updates for their end-user terminations?
- What are the minimum term implications for end-user terminations?

The high-level system functions for end-user service termination/cancelation include:

- validate termination/cancelation request data;
- process internal service termination and cancelation;
- process service termination/deactivation exceptions;
- update billing.

8.8.7.1 Validate Termination/Cancelation Request Data

The following functions are to be considered:

- validate data integrity with acknowledgements;
- validate if end-user requests already exist (i.e. check for duplicate requests);
- simple field/order information validation (name, CLI (if any), address, post code, etc.);
- if orders are in batches, separate batch orders into individual end-user orders/cease/cancelation requests with acknowledgement to each end-user order;
- there is a need to release the network resources after an end-user's service has been terminated.

8.8.7.2 Process Internal Service Termination and Cancelation

Functions to support the customer network termination and cancelation may include:

- deprovision customer on access and core network according to network resource specified;
  - ○ ensure that the network is deprovisioned on the specified date;
- cancel the relevant service provisioning order;
- deactivate the end-user's service profile;
- update network inventory to release network resources;
- update billing system.

8.8.8 Process Service Termination/Deactivation Exceptions

There may be instances where service termination/cancelation and deactivation activities may fail. These failures and exceptions are dealt with by operational personnel (normally system users).

As part of the system design activities, a list of possible exceptions should be stated and an operational process will be required to deal with the exceptions from the system. Below are some system functions that will help with order rejects and exceptions:

- identify the service provisioning and deactivation and group with reject reasons;
- update the termination/cancelation requests with reject reasons;
- automatically fix exceptions within system where possible;
- generate report for manual interventions;
- GUI for system users allowing them to edit requests (both internal and other network operator orders).

8.8.9 *Update Billing*

Before billing is activated, there should be some testing performed on the end-users' service to ensure the service and network provisioning activities are successful. Depending on the service, this may only be a simple ping test, as it is difficult to test whether the service profile is working until the end users use the service. The other choice is to call the end users to ensure that they are happy with their new service. This could prove a bit costly. Again, it depends on the kind of service you are providing.

8.9 Billing, Charging and Rating

Billing is one of the most important elements within a service. Without the mechanism to charge the customers/end users accurately, you are at risk of losing revenue and potentially going bankrupt. However, quite often, we do not think about billing until the end, as most people think it is easy. It is not unusual to put in place a manual billing solution before the service grows to a certain size or a new feature is put in place in the current billing system, as it cannot generate the bill required. It is also not surprising that some companies do billing manually for a service. I cannot comment on the accuracy of manual billing as a solution, as this is highly dependent on the caliber of the people creating the bill. However, manual billing is not recommended, as it is not sustainable.

When designing the billing solution, one may like to ask the following questions:

- How many customers and/or end users are going to be billed?
- What is the rate at which new customers and/or end users are added to the service?
- Where will the usage data come from for usage-based charging items?
- What is the estimated number of usage records per day per customer/end user?
- Is this service sold as a bundled service with another (e.g. phone, broadband and cable TV service as one package)?
- What format will the bill be in?
- Do the customers expect a paper bill or electronic one?
- What are the billing fulfillment requirements?
- Do all the accounts have the same bill date?
- How will the bill payment be accounted for?
- Is there a requirement to track overdue payment?
- What are the bill payment methods?

The principle functions within the billing area are:

- defining service charging structure;
- creating and terminating customers/end users on the billing system;

- billing data collection;
- rating and charging;
- invoice creation and dispatch;
- bill payment collection;
- supporting bill enquiries and disputes.

8.9.1 Defining Service Charging Structure

In order to define a service on a billing system,[1] the first thing we need to work out is the service charging structure. This should have been defined by the service definition that forms part of the business case.

Most telecom services are broadly made up of the following charging elements:

- one-off installation/sign-up charges;
- recurring service rental;
- usage-based charges – this can be time based (e.g. call charges of $x per minute) or quantity based (e.g. $x per Mbit download of content or charging $y per piece of content);
- quality based – this is mainly used for QoS-based charging;
- call plan and service package or discount packages;
- billing and rating period of all the above items.

Most service contracts will have a minimum term. The penalty of early service termination will need to be defined such that penalty charges can be charged accordingly.

Depending on the flexibility of the billing system used, one needs to bear in mind that there might be items that are billed or discounted for customer special requests. For example, the customer may want a dedicated operational person to service their provisioning request and are willing to pay extra for that service. We need to be able to cater for this type of scenario. The implementation of this depends on the flexibility of the billing system.

When designing the format and style of the bill, one needs to bear in mind how this service fits with other services being billed. One customer/end user may buy multiple services from one service provider.

The billing cycle and charging periods of all the chargeable items need to be set. Some customers may have specific requirements for when they would like to receive the bill, as they might want to manage their cash flow. The billing cycle defines when the bills are generated for which service and for which portion of the customers within that service. Depending on the design of the service and the billing system, billing cycles for different services may be different. It is not always efficient (and can be very expensive) for the billing system to rate and generate bills for all the customers/end users at the end of each month. The load should be spread more evenly over a period of time. Not only does this even the load on the system, but this also helps with monitoring and resolving billing errors. If there is a problem with billing a portion of the customer for a particular service in a billing cycle, then, hopefully, the problem will be resolved before the next billing cycle or bill run where the next batch of bills are to be generated. The rest of the customers will not be affected. If the problem is not resolved when the next bill run is due, then the billing management team will need to decide if the next bill run should take place and the risks involved in doing so.

---

[1] When writing this section, I am assuming that there is a billing system already in place. Procurement of billing systems is outside the scope of this book. However, the system functions and considerations described in this section will certainly give you some useful pointers.

8.9.1.1 Fixed-Line Services

Time-based charges can also be distance related. For example, the rate for a local call is different from a national call, and international calls may be of different rates for different countries. When setting up the billing structure and rating tables for the calls, there will be a requirement to distinguish different charge rates for different types of call. In addition, there may be different charges for time-of-day usage. For example, calls in the evenings and weekends are cheaper than those during the week and during the day. Therefore, the rating tables need to be set up to cater for this requirement. Freephone/toll-free (0800/1-800) numbers and premium-rate numbers (0870, 09xx) are all charged differently. All these need to be defined in the structure before different rates can be entered into the rating tables within the billing system.

8.9.1.2 Mobile Services

From a mobile carrier's point of view, all the fixed-line principles apply. However, additional areas for consideration include:

- Different rates are applied to calls made to the mobile phone in different countries, as different charging arrangements and call rates exist between foreign operators.
- CDRs associated with roaming calls may also be of a different format. These CDRs may come through a data clearing warehouse, where the CDRs are reformatted before passing through to the relevant mobile carrier.
- For a particular call or session, the end users will be using multiple network elements for a call/ session while moving from one location to another and at the same time using different applications. Hence, call information can be coming from multiple network elements in the access and core network, the session servers for the session control and the application servers for the use of application. Therefore, usage record correlation can be complicated. In addition, these record formats are unlikely to be consistent; therefore, there is a requirement for a mediation system to ensure consistency of data and format before input is fed into the billing system.
- Owing to the diversity of applications and services on offer, new categories of usage (e.g. value of content rather than time base usage or data volume) are to be supported. With diversity of services and application and content provision, the supply chain and payment settlement between different service providers and third-party suppliers/content providers within the chain can be complicated. There may be requirements to support the different revenue apportionment/revenue share arrangements for the different parties in the chain. The accounting of this will also need to be managed. A clearing-house arrangement may be required.
- The payment method for bills, services and applications may include real-time/online credit card or micropayment methods, especially for event charging functions (e.g. purchase of content), which require third-party authorization. Therefore, a real-time charging and bill update mechanism is needed (rather than just conventional monthly bill payments or prepaid mechanisms/arrangements).

8.9.2 Creating and Terminating Customers/End Users on Billing System

Creating customer/end-user billing accounts enable the system to identify who and where the bill needs to be sent to. As stated in Section 8.3.1, customer accounts are created when they order the service. It may or may not be necessary to create end-user accounts, depending on the billing relationship with the end user (i.e. if the end users are billed directly from the system). Below are some system functions to consider for creating and terminating customers/end users for billing.

- Create customer/end-user account details, including billing address and customer contact person regarding payment chasing. (This not relevant for existing customers.)
- Assign billing items bought by customers/end users.
- For customers/end users terminating/ceasing the service, change the status to cease (e.g. from active to cease). It is a good idea to keep customers'/end-users' billing records for any overdue payments issues.
- Input percentage and amount for volume discount parameters (if relevant and if different from standard charges).
- Input price variations on billing items (if different from standard charges of service; appropriate authorization is required).
- Input minimum term for termination (if different from standard service).
- Define billing cycle for the customer (e.g. Do they have monthly bills or are they billed quarterly? Is there a specific date the customer wished to be billed on to manage their cash flow?)
- Input penalty for early termination (if different from standard).
- Select electronic or paper bills;
- For account termination, check for any outstanding or termination/cancelation charges.
- Define payment methods (direct debit/check, etc.) and payment terms (i.e. when the payment is due). If payment method is direct debit, then the system function may include setting up of direct debit and holding end-user bank account details.

8.9.3 Billing Data Collection, Correlation and Mediation

In order to rate the billing items correctly, the data to be billed must be collected. For usage-based charges (e.g. voice calls), usage data is generated from the network equipment. For example, for PSTN calls, CDRs are generated from the voice/circuit switches after the call is completed. CDRs contain data like call start and end time, total call duration, calling party number and number called.

As part of billing data collection, mediation between different sources of data may be required to ensure consistency of information and data format. The functions for a mediation system include:

- data collection from multiple sources (e.g. extract/download usage data from the network element daily, depending on the size of date);
- verify all required data is present and data is not corrupted;
- apply filtering rules for data and automatically correct any data error in accordance with business rules where possible; otherwise, exceptions should be raised;
- identification, aggregation and correlation of different data records for one billing transaction;
- reformatting of data collected to the format that is acceptable to the billing engine;
- store data records for billing engine to produce invoices.

The usage data will be required when generating the bill and for customer/end-user bill enquiries. For some countries, these call records are to be stored for a longer period of time owing to local regulations For example, in the United Kingdom, the regulatory requirements for storing CDRs is 7 years. Depending on the service and volume of data involved, one needs to decide whether this data is to be imported into or exported out of the billing system. A mechanism to view the bill and the usage records will be required for bill enquiry purposes.

8.9.4 Rating and Charging

Rating and charging ensure that all the chargeable items are rated correctly and that the customers/ end users are charged the correct amount. Functions to support this task include:

- process and manipulate billing data (e.g. CDRs for voice service) according to business needs and business rules;

- charge customer new installations (if any);
- charge end-user new installations (if any);
- rate and charge line rentals to customer sites (if any);
- rate end-user rental (different end-user service packages many have different charges);
- rate usage charges – this could be a time-based or quantity-based, and rates may vary according to distance (e.g. national versus international calls, local calls versus national calls);
- charge any service-related items (e.g. reports, special requests/services);
- apply service package discounts/call plans (e.g. inclusive number of minutes per month or package discount on phone calls with broadband service);
- service rental with percentage discount over specified volume – different discounts may be applied to different customers at different tiered rates for different volume break points;
- check for minimum term for cancelation (e.g. if canceled/terminated within the minimum term, then the customer/end user may be charged for the remainder of minimum term or a fixed cancelation charge);
- rating of calls by distance (e.g. international calls with zones defined; national or local calls);
- provide the facility to input service credits, rebates or promotion discount due to the customers/end users;

  o this is normally a manual task, as service credits, rebates or promotion discounts need to be authorized; for service credits, they are only paid if SLA breaches have been agreed between the customers/end users and the operator.

- apply relevant taxes to the bill where chargeable (e.g. VAT);
- support of multiple currencies should also be a consideration.

8.9.5 Invoice Creation and Dispatch

Once all the billing items have been rated, bills are ready to be created and dispatched.

8.9.5.1 Paper Bills

Paper bills need to be printed and sent to the customer/end users. Most operators outsource this function. A predefined formatted file will be sent to the outsourced agency for printing.

8.9.5.2 Electronic Bills

If electronic billing is a requirement, then one needs to decide the mechanism of delivery. Depending on the requirement, it could be e-mail, ftp to customer server or present it on the reporting portal for customers to download. All security considerations need to be assessed for various access mechanisms.

8.9.6 Bill Payment Collection

For payment collection, the system functions may include:

- Collect bill payments. The system may need to support the collection of payment via direct debit, record check payment and interbank transfers, and so on.
- Tracking of bill payments from the customers/end users. This may include tracking which customers have paid and the amount paid.

- Update payment status for each customer/end user needs to be updated once the payment has been received.
- Check for overdue payment and send payment reminders where relevant.
- Notify system users of customers'/end-users' overdue payment/debt for a specified period of time.

8.9.7 Support Bill Enquiries and Disputes

System functionality to support billing enquiries includes:

- Providing the facilities for the bill enquiry team to access customer/end-user bills and billing records. This may include CDRs or customers'/end-users' log-in session details.
- Logging of customer/end-user enquiries and billing disputes. This includes information like the nature of the enquiry and answers/resolution to enquiries/disputes.
- Providing facilities for the billing operational team/billing enquiry to adjust customers'/end-users' bills following disputes that have been resolved.
- Monitoring and ensuring that all enquiries are dealt with within the customer SLA.
- Monitoring the number of outstanding billing enquiries.
- Providing reports on the performance of the bill enquiry process. Example reports may include:

  o number of bill enquiries per week;
  o average bill enquiries' resolution time;
  o number of outstanding enquiries;
  o age of outstanding bill enquiries;
  o number of bill enquiry escalations per week/per month.

8.9.8 Billing, Charging and Rating Summary

A summary of the billing, charging and rating functions is provided in Figure 8.9.

8.10 Service Accounting, Revenue Reporting, OLO Bill Reconciliation and Revenue Assurance

8.10.1 Service Accounting and Revenue Reporting

Service and revenue accounting requirements are often not apparent because most people assume these will 'just happen'! Some finance departments have separate financial systems and system support that are not transparent to the rest of the company. As a result, their requirements are often forgotten or not dealt with when new services are introduced.

Most product managers have revenue targets for their products/services each year. Without a tool to perform the accounting function, it is often done manually or not done at all. Accounting the revenue manually may be cost effective if the service is straightforward and if the information is easily obtainable. However, how do we know if the information is easily obtainable without investigating the requirements? Therefore, the following questions are worth asking, to decide whether any system development is required in this area:

- Is there a requirement for accounting for the revenue for this service?
- How does this service fit within the existing service accounting framework?
- How is this service different from others?

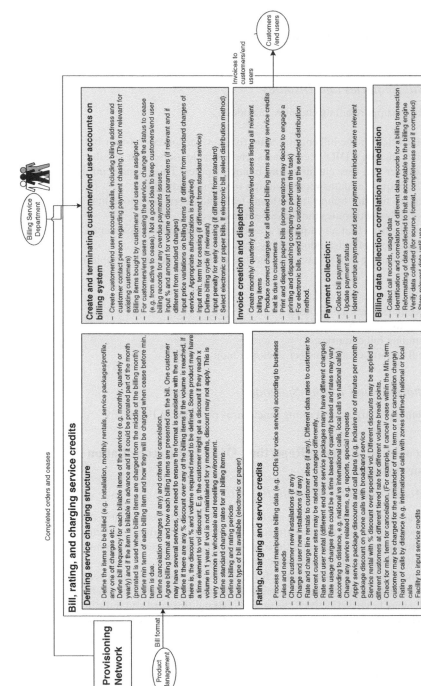

Figure 8.9 Billing, charging and rating summary.

- Will the revenue be accounted on a per customer basis?
- Where will the accounting information come from?
- What is the accounting period and frequency of reports?
- What are the additional requirements for this service (assuming this function already exists)?
- If it is a bundled service, how will the revenue be accounted for/split between the various services that make up the bundle?

If it has been decided that service accounting and revenue reporting are required, then you may want to consider the following reports:

- Number of new customers/end users this month.
- Number of termination before minimum term is due.

  o It may be useful to group this by customers.

- Revenue per month.
- Revenue per month by service.
- Revenue per month, broken down by customers.
- Overdue payment.

  o A list of accounts and the amount due and due date will be useful. The reports may be grouped by due dates (e.g. accounts overdue < 30 days, 30–60 days) or by the amount due (e.g. < £500, £501–1000).

- Account for payments to OLOs (including roaming charges for mobile calls).
- Account for payment due for third-party service providers/content providers/application providers if the end-users'/customers' payments include them in the value chain as part of the service.
- Report to enable the calculation of OLO cost per month.
- Assist generation of report for customer and service profitability.

  o Profitability reports on a per service basis are always difficult since the cost per service is not always clear and accounted for, as resources and costs are often shared between services to achieve cost and operational efficiency.

8.10.2 OLO Bill Reconciliation

Bill reconciliation in this context is reconciling charges from other operators (OLOs) you are buying a service from or roaming charges for mobile calls. This is to ensure you are not paying more than is due. This function applies to all services that use external operator services. In the case of mobile services, this may mean roaming charges and payment due for third-party content/service providers. At the start of the development, one needs to ask:

- Which bills need reconciliation? What chargeable items or transactions require reconciliation?
- Where will the data come from and in what format (e.g. paper or electronic file)? If electronic file, what format will the file be in?
- Is there a system that performs the current function?
- Will this service be reconciled the same way as other services?
- What are the additional items needs reconciling?
- How often will the OLO bills arrive?
- How often does the bill reconciliation need to happen?
- What is the estimated volume of data?

- How long does the data need to be kept for?
- Is this operation expected to be a fully automated or a semi-automated one?
- Who will be the users of the system?

When designing this part of the service, one needs to investigate how the company is doing it currently and what additional provisions are required for the service you are designing. The development may be minimal, as the current framework might incorporate the new service very easily. If new billing items are to be included, then development may be required.

Below are some examples on the different types of charges that need reconciling

- normal calls between operators;
- freephone/toll-free calls (0800/1-800);
- premium-rate calls (0845, 0870, 09xx);
- mobile roaming charges (both calls and SMS);
- lease circuit from OLOs;
- wholesale products/connections bought from OLOs (e.g. broadband end-user connections).

The system function for this area is relatively simple: reconcile the bills. However, when defining the system functions, you may want to consider:

- Defining the billing items to be reconciled;
- How to rate each of the billing items (amount that should be charged per billing item);
- Any pro-rated charges involved;
- Define tolerance level before generating an exception.

    o There can be lots of exceptions due to rounding differences or clock timings. To minimize unnecessary exception and improve efficiency, a defined tolerance level should be agreed such that only real exceptions are listed in the exception reports.

- Generating a list of exceptions/exception reports for items that cannot be reconciled or the variance exceeds the tolerance level.

Data on which the reconciliation is based will come from the network for usage-based items (e.g. call records (CDRs) for voice calls) and from the network inventory (or equivalent) for flat-rate-based (or bandwidth-based) items. Identification of a reliable/definitive source of this information (especially for flat-rate items) and the mechanism to extract these items from the appropriate places will need careful consideration.

8.10.3 Revenue Assurance and Fraud Detection

Revenue assurance is a topic that is often forgotten, but it is important to ensure that the revenue to be collected is accurate and that the revenue due is collected. Another element to revenue assurance is fraud detection. When considering the system functions in this area, the following questions might help:

- Are there any revenue assurance requirements?
- Where do revenue leakages occur?
- How much revenue are we trying to protect?
- Are the revenue assurance activities manual?
- What are the revenue assurance requirements on the systems?
- What mechanisms are required to prevent revenue leakage and potential fraud activities?
- What additional reports are required?

It is virtually impossible to check every single CDR or bill produced for accuracy. However, some system functions should be put in place to ensure relevant checkpoints have been set up to minimize revenue leakage. Below are some system functions/measures, reports and checks that can help with revenue assurance and minimize fraud opportunities.

- Reconcile the actual activated customers/end users in the network with customers/end users that are active in the customer management system and billing system (plus other systems that may hold customer/end-user status). This can reduce fraudulent use of the service.
- Monitor the revenue billed per month. Drastic changes normally indicate billing error, unless special promotions or events are taking place that month.
- Reconciling bills from other operators with internal CDRs as stated in the previous section. This is especially important for roaming charges for mobile services.
- Monitor the CDR volume and the number of records that are rejected by the system during processing and manipulation. A high number of errors and regular failures will indicate problems in the billing chain. This can cause potential loss of revenue.
- Monitor bill rating errors. Rating errors will also cause revenue leakages.
- Perform random checks on bill content to ensure bill accuracy. Test accounts, calls and data transfer can also be used for verification purposes.
- Cross-checks between CDRs and network usage data from network performance reports. The estimated usage of the network can be obtained from the switches. The usage data should be broken down into calls for different call types (e.g. mobile to fixed line, calls going to/received from other operators or mobile-to-mobile calls). Comparing the total usage of each call type for the same time period taken from the CDR data and the network performance data will provide a good indication of potential revenue leakage for voice calls. Similarly, for usage base data sessions, the network usage data from data switches or routers is used to compare with the amount billed to the customers/end users.
- Monitor unusual usage patterns. An unusual usage pattern can be an indication of fraudulent use of service.
- Monitor system and network outage (both internal and interconnect networks to other carriers). System and network outage can cause potential call failures and data nondelivery and, hence, potential loss in revenue. For calls to be delivered to or received from other operators (OLOs), one might want network performance reports to ensure the interconnecting operator is fulfilling the contractual SLA on network availability and service outage notice agreements to avoid potential loss of revenue for both parties.
- Monitor overdue payments. Overdue payment reports should be generated for all customers and all services. To avoid bad debts, it is important to carry out credit checks on all customers and end users. Encouraging customers/end users to set up a direct debit is a good way to collect revenue and avoid bad debts.

The perception of most people is that revenue assurance is a function that resides within the finance department. However, as demonstrated above, these are cross-functional tasks, though the responsibilities do lie within the finance department.

8.10.4 Service Accounting, Revenue Reporting, OLO Bill Reconciliation and Revenue Assurance Summary

A summary of the service accounting, revenue reporting and OLO billing reconciliation functions is provided in Figure 8.10.

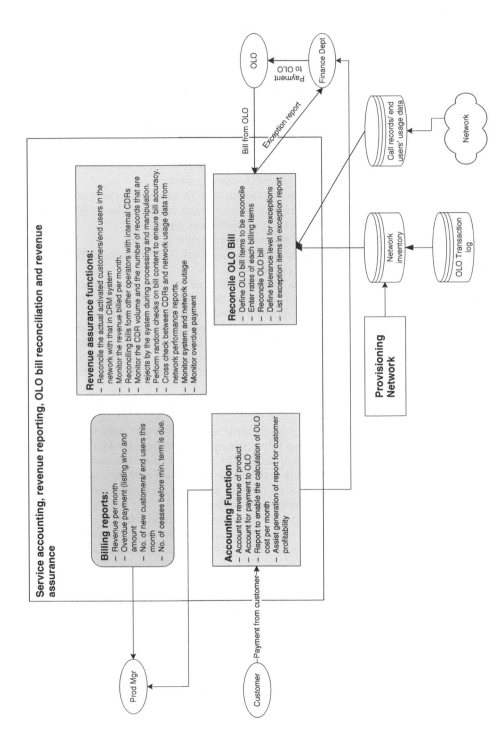

Figure 8.10 Service accounting, revenue reporting and OLO bill reconciliation summary.

8.11 Fault Management

Fault management in the context of this book is managing faults that are raised by customers/end users. Faults raised by the customers/end users can be caused by the network and failures in the support systems. The system functions for this area will need to support both the customers' requirements and internal requirements from operational areas. Questions to be asked when gathering requirements for this area may include:

- Who can raise fault tickets?
- Where will fault tickets come from?
- What are the mechanisms by which the customer/end users can raise a fault against the service (e.g. by ringing help desks, report them through a portal)?
- What eternal interfaces are required for the system?
- What internal interfaces are required for the system?
- What needs to be done to qualify a fault?
- Will there be a requirement for a diagnostic tool for customers/end users to use before faults can be logged?
- Will there be any diagnostic questions for the customer/end users to answer to improve fault diagnostic accuracy/efficiency? If so, what are the questions?
- Is there a requirement to develop remote diagnostic tools for internal users to aid fault diagnosis?
- Will the faults be categorized? If so, how many categories and what are the criteria and the associated SLA for each category?
- Will different customers have different SLAs?
- Do the customers/end users expect fault status updates? If so, what do the updates consists of?
- Are the status updates sent proactively to customers/end users when there is a status change, or is the information to be made available for the customers/end users to view? Or does the system need to provide the facility for the help desk to call the customers with the fault status updates?
- How will the existing network and system faults be linked to faults raised by customers/end users?
- How do you inform customers/end users about known faults to reduce the volume of duplicate fault tickets?
- What is the estimated volume of faults and their arrival rate?
- How many simultaneous system users does the system need to support?
- What are the nonfunctional requirements for the fault management system (e.g. response times to perform requests and at what load level)?

The rest of this section highlights the functions for consideration when developing a fault management system/solution.

8.11.1 Fault Ticket Validation

Before a fault can be logged onto the system, validation needs to take place. The following functions should be considered:

- Enable fault ticket input (file or from GUI).
  - Depending on the service, customers/end users may report a fault through a help desk. So, a GUI will be required. If customers/end users report the fault over the Internet or system interface to the customer's system, then a system-type interface with a Web-based front end will be required.
  - To qualify a fault, some initial fault diagnostic questions will be required to ensure the fault is genuine.

- Validate the fault ticket or that the file containing the fault tickets is coming from an expected source.
- Check there is no duplicate fault ticket.
- Simple field information validation to ensure that all the fields are completed in the correct format.
- Generate fault ticket reference.
- Assign fault severity (if possible).

    o Through the fault diagnostic questions, the system may be able to assign a fault severity depending on how the diagnostic questions have been set out.

8.11.2 Fault Identification, Log and Track Faults

Once the fault has been logged, the system should, where possible, perform an initial diagnostic. For example, if the fault is caused by a planned outage, or by any known faults or any faults that have been reported by OLOs, then the status of the fault ticket should be updated with the expected resolution time. If the system cannot find an existing fault that has been logged, then human attention will be required. All actions performed on the fault tickets should be logged and tracked.

8.11.3 Fault Diagnosis and Fix Fault

Fault diagnosis and fixing faults are normally performed by human actions. From a system point of view, the following functions should be considered to assist the human actions:

- correlate fault reports and group by geographic areas or node locations;
- perform fault correlation where possible;
- isolate faults and perform root-cause analysis where possible;
- log fault diagnostic and resolution.

8.11.4 OLO Fault Logging and Updates

If it has been identified that the network connection provided by the OLO is at fault, then a fault is raised to the appropriate OLO. All these faults are to be logged and the SLA measured.

- create fault tickets with OLO with interoperator fault ticket references;
- obtain fault ticket updates from OLO;
- update internal fault tickets with OLO updates;
- alert system user if OLO fault resolution is 'no fault found';
- track OLO SLA and alert system users of open fault tickets that are about to breach the SLA with OLO.

The above assumes that the OLO has an automated fault logging system.

8.11.5 Work Flow and Jeopardy Management

As part of the fault management function, the system needs to ensure that the fault is resolved within the customer/end user SLA and that all tasks are performed by the correct fault resolution agency. The following functions should be performed:

- update status of fault tickets;
- ensure tasks are performed by the correct resolving agency and track progress;
- alert system users for fault tickets that are about to breach given SLA or carry out any specified escalation procedures.

8.11.6 Update Fault Status with Customers/End Users

To provide good customer service, regular updates on fault status and fault resolution are essential. Therefore, the system functions to assist this should include:

- Sending fault ticket acknowledgements for new fault tickets.
- Provide fault status with time stamps.
- Provide the facility for the customers/end users to view fault ticket status.

  o The mechanism for this depends on the method of access. For business customers, this may be a system interface, in which case the system needs to track and send the latest updates.
  o The fault status update may include action required by the customers or end users to resolve the fault.

Fault status could be:

- open;
- no fault found;
- await OLO feedback;
- await further information from customers/end users for further diagnostic;
- ongoing investigation;
- closed.

All the status updates should be date and time stamped.

8.11.7 Fault Management Summary

A summary of the fault management functions is provided in Figure 8.11.

8.12 Network Management (Monitoring and Collecting Events from the Network) and Service Management

In order for the customer to have a good service experience, it is essential to ensure that the network that the service is reliant on is functioning as expected and that faults or degradation in service levels are noticed and fixed at the earliest opportunity. Effective network management not only helps to maintain customer SLAs, but also ensures a positive customer experience.

Network management in the context of this book is the monitoring of network status, collecting events from network elements and correlation of network events. For those who are familiar with the telecommunications management network model, this section of the book covers the functions in both the network management layer and the element management layer. The functions listed in this section may be split between the element management systems and network management systems, depending on the implementation of these functions and the system architecture of your company.

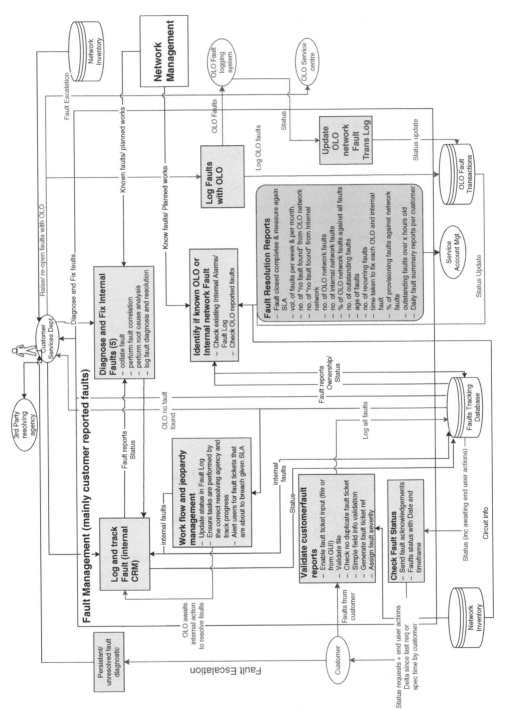

Figure 8.11 Fault management summary.

8.12.1 Network Management Requirements

When designing the network management system, the following questions should be considered:

- What are the communication protocols between the network device and the network/element management system?
- How many network elements are being managed and monitored?
- What is the average number of interfaces per network element monitored?
- What are the network parameters to be collected?
- What are the network protocols to be used for the communication between the network management system and the network elements?
- How many parameters are to be collected per network interface and network element?
- How frequent will the network element/interfaces be polled for information?
- What are the nonfunctional requirements for the network management system (e.g. the response time required in processing network events and alarms, updating the network status)?
- What is the network management architecture for the service? Distributed, centralized or hierarchical?
- Where will the network management data be stored?
- What is the estimated size of data over a month and a year?
- How long does the network management data need to be stored for?
- Which network management parameters require archiving off to a separate location and which ones are stored locally on the system?

The principal system functional areas for network management include:

- monitoring network status and collecting network events;
- alarm correlation and troubleshooting;
- performing troubleshooting on the network;
- logging internal network and OLO network faults;
- configuring network parameters and configuration management;
- correlation between network assurance, service assurance and service management;
- network inventory.

8.12.2 Monitor Network Status and Events

8.12.2.1 Monitor and Logging Network Status

Monitoring network status involves proactively gathering network events from the network. This involves polling the network elements to obtain network status information. The major functions for monitoring the network status are:

- define the network parameters/characteristics to be polled;
- define the network elements and the network interfaces to be polled;
- define polling intervals;
- poll the network interfaces for data;
- collect polling data from network elements with date and time stamp;
- validate data to ensure information is complete, genuine and uncorrupted;
- store network polling data;
- define network parameter thresholds for network events to be raised;
- raise and log necessary network events.

8.12.2.2 Monitor and Logging Network Notification/Events

Network elements send out network notifications to indicate problems that have occurred and potential problems that may occur. These notifications can be notification of problems or potential problems or failures in network elements or certain preset thresholds or conditions have been reached or exceeded.

The major functions for monitoring the network events are:

- Collect network status information/network notification (i.e. SNMP traps).
- Validate data to ensure information is complete, genuine and uncorrupted.
- Log network events.
- Monitor repeat faults and recurring problems.

  o Repeated faults indicate that the original network problems have not been resolved. Further diagnostics and actions will be required.

Network events can include: link up/link down, interface up/down power warning, I/O error, unable to open port, etc.

8.12.2.3 Filter and Process Network Notifications/Traps and Network Events

The following functions are carried out as part of the filter and process network notification:

- define network event filters and filter configuration;
- apply filters to network traps (the filters can suppress notification, as a faulty network element can cause too many notifications);

  o user filters may be applied as each system user may only want a certain view of the network that they are responsible for);

- translate network traps to network events;

  o translate SNMP traps (using the definition in the management information base (MIB)) to a more meaningful event type;

- generate network event if characteristics exceeds predefined thresholds/conditions;
- forward the network event to defined system or person.

8.12.3 Alarm Correlation and Distribution and Trouble Shooting

One other important functional area within the network management system is alarm correlation. The system needs to provide functionalities to help the system users (normally network management personnel) to diagnose the network faults. Alarm correlation is an effective way to help fault diagnostics. When designing the system, one may consider the following functionalities:

- categorize the severity of events into alarms and alerts;
- define the format of graphical display of alarms;

  o most 'critical' alarms are labeled red, such that the alarms can be spotted easily; other are up to the users to define;

- define alarm groupings (e.g. based on network topology/network inventory information);
- define alarm filtering rules (e.g. duplicate alarms, auto-clearing rules) to reduce the number of unnecessary alarms;

- define rules for alarm correlation;
- group alarms according to groupings or pair up events to indicate source of fault;
- filter alarms in accordance with the filtering rules defined;
- correlate events and isolate faults where possible;
- associate alarms with service where possible;
- display of alarms in the defined grouping and categorization;
- provide GUIs for the system user to drill down/access to the desired network elements for additional historical information for troubleshooting and fault isolation;
- view network logs/historical events and details of network elements.

Alarm severity categories can be defined as follows to help network managers to prioritize work:

- 'critical' – require immediate attention;
- 'major' – actions required as soon as possible;
- 'minor' – potential service degradation, monitoring required; or
- 'warning' – potential problem may occur, investigation required.

One may also want to mark the faulty network element in the network inventory. This may be useful for network provisioning and fault management systems/functions to identify a faulty network element.

8.12.4 Perform Troubleshooting on the Network

Troubleshooting on the network is normally performed by network managers. System functions that can aid with the tasks include:

- displaying status of network elements and their interfaces;
- displaying event history for network elements and their interfaces;
- facilitating communication with the network elements directly (e.g. providing a command line interface (CLI)).

8.12.5 Log Internal Network and OLO Network Faults

All faults should be logged in the trouble ticketing/fault management system to ensure they are resolved within the agreed SLA and that any workflow and jeopardy management mechanisms are put in place. If faults have been identified to be with the OLO's network, then there will be requirements to raise and log fault tickets with the relevant OLO. Monitoring the status of OLO fault tickets will also be required to ensure they are resolved within the SLA.

8.12.5.1 Logging OLO Major Outage and Planned Works (If Required)

For services relying on network resources from another network operator (OLO), it will be useful to log all known major service outages and planned works to minimize unnecessary fault tickets and to help with fault diagnostics.

8.12.6 Configure Network Parameters and Network Configuration Management

Configuration management functions are used during the installation and configuration of network elements and recovery of network elements after they have been restored from failures.

Configuration management functions also provide the system users with the means to control and manage the configuration of the network. The system functionalities may include:

- maintaining the end-to-end network configuration database/network inventory;
- controlling network connectivity and routing changes;
- recording network elements additions/deletions;
- downloading configurations to network elements (using FTP/trivial FTP);
- uploading network element configuration in a predefined period;
- storing and backing up network equipment configuration;
- defining configuration template for each type of network equipment;
- applying relevant configuration templates during installation or network provisioning;
- ability to roll back to previous configuration;
- ability to download new software version to the relevant network elements;
- conducting audit on configuration of network elements and network inventory.

Configuration management mechanisms may include:

- direct access to the network elements;
- remote access via the CLI or telnet.

8.12.7 Correlation Between Network Assurance, Service Assurance and Service Management Functions

Having network events and alarms is not often useful from a service management point of view. If a certain piece of network equipment has a fault, what does it mean to the service? What service is affected by this network fault?

The correlation between the network events and the service affected will deliver the true meaning of service assurance. When designing this functional area, one needs to consider which approach is most appropriate. Do you have a 'bottom-up' approach looking at the network and relate each network element to the services and customers it supports? Or do you use a top-down approach from the service viewpoints, identifying which network elements are related to the service and assess the service impact of the customers when one of the network elements has an event? In an ideal world, you like to have both viewpoints. The former will be more useful for the network management center to fix the problem, while the latter will help the service managers/account manager to manage their customers.

Therefore, each network element will have 'relationships' with multiple services, while a service will have relationships with multiple network elements. Mapping the network elements to a service is essential. This relies on the service being marked on the network element or vice versa. The implementation of this function depends on the system architecture in place. The network inventory (as described below) can be a good place to hold the information. The mapping of this could be done as part of the service provisioning task.

8.12.7.1 Service Management

Service management is generally performed by operational teams (rather than by operational systems) with the defined service management process. However, system functions to help with service management may include:

- mapping of services to network elements;
- assessing service impact as a result of network faults;
- analyzing service performance from network performance data (e.g. data error rate, network throughput data from network interfaces);

- ensuring SLAs are met and issue a warning where SLAs are about to be breached;
- ensuring both internal and customer service and network performance reports are established, scheduled and delivered;
- making relevant service data available for view by customers/customer reporting systems;
- performing service impacts on planned work and major service outages;
- informing customers of planned work and planned service outages, which can be an automated process if service impacts are performed by the system.

Where SLAs have been breached/violated, processes to determine the information to be delivered and the delivery of the SLA violation information to the customers will be required.

8.12.8 Network Inventory

Having an accurate inventory is essential to effective network management. The network inventory should track the physical location, physical configuration of network elements (e.g. card, slots, shelf), physical connectivities (e.g. copper, fiber), functional equipment (e.g. interfaces, channel) and logical connections and available capacity (e.g. E1, STM-1) of all network elements. From the network inventory, an accurate network map can be drawn giving an overview of the status of the network. It can also be used to manage spare and faulty parts.

Creating a network map to represent the interrelationships of various network elements and network connections and interfaces will be a useful tool for impact assessment resulting from network incidents. Automatically discovering and mapping of network elements could be part of the network inventory functions. Updating the network inventory should form part of the network element installation and commissioning process. However, human error does occur and so the autodiscovery functionality may be preferred, as this automatically synchronizes between the network (including network elements and their constitute components, e.g. network interfaces) and the information held in the network inventory database.

The network inventory may also store information that is related to the network element. For example, model number, physical location, current configuration, software versions, physical connections (including the connection types and capacity of connections), IP addresses, current fault against each network element, past fault history, last upgrade, capacity availability, service supported and so on. The information held in the network inventory may also help with network provisioning activities for the customers/end users.

For details on network inventory data model, please refer to *OSS Essentials: Support System Solutions for Service Providers* [53].

8.12.8.1 3G Mobile Network Inventory

For a 3G mobile network, additional data in the network inventory may include:

- network node types (SGSN, GGSN, RAN, MSC);
- physical link (e.g. microwave, fiber, copper);
- core network domains (e.g. CS domain, PS domain, IMS);
- remote site locations.

The number of addressable devices in 3G mobile is forever increasing in number and complexity; therefore, the inventory management of a 3G mobile network is going to be challenging.

8.12.9 Network Management Function Summary

A summary of the network management functions is provided in Figure 8.12.

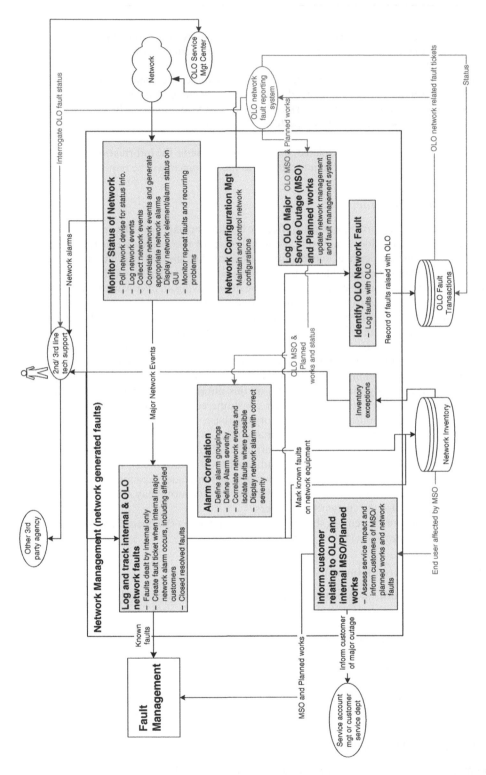

Figure 8.12 Network management summary.

8.13 Performance Management

What is performance management? Performance is the measure of how well the system/application and network are doing. It involves the measure of: how many tasks or functions it can perform at any one time; how responsive it is (time taken) to certain conditions or events; how often certain conditions may occur; how available the system/network is for the service; and what the tolerance levels are to certain conditions.

Performance management, from the systems functions point of view, mainly involves gathering data from the network elements, systems or application being monitored to ensure they are performing as efficiently as possible. When designing a performance management system/tool, the actual performance of the tool is often forgotten and nonfunctional requirements for the system are overlooked. The performance of the performance reporting tools is just as important as the performance information to be gathered and analyzed. The following questions are designed to help you to think through the requirements for the performance management system and the reports to be considered when measuring network, system and application performances.

With regard to *network performance*, questions to consider include:

- What are the network elements and their interfaces that require performance monitoring?
- What are the goals for network performance? This dictates the directions of the tools to be used.
- What are the overall network performance targets?
- How does the network performance translate to service performance? What are the relationships between the service performance and network performance? Which of the service performance measures (SLA, service KPIs) are based on network performance?
- What are the network performance measurements required to meet the performance targets?
- Will averaging data be meaningful?
- What are the parameters to be measured? Will round-trip delays, packet loss and jitter statistics suffice?
- What are the main sources of data in terms of network elements (routers/switches/CPE), links, interfaces, traffic routes, traffic source and end points?
- What are the traffic routing considerations? What routing policies are used in the transmission and routed (layer 1 and layer 2) network? What will be the performance impact under failure scenarios?
- What is the traffic mix for the service? How will it affect the performance of the network? Will this need to be modeled in the performance management tool?
- What are the different types of traffic that require performance measures?
- How will the network perform under various failure scenarios (e.g. With a network node going down, what is the impact on the network while traffic is being rerouted? What will be network performance impact after the traffic has been rerouted?) Is simulation of the rerouting scenario a requirement for the performance management tool?
- What types of traffic simulation are required?
- Where do we get the traffic data?
- What is the mechanism/protocol used to retrieve performance data from the network?
- What is the frequency for data collection?
- Where are the points at which network performance is measured? Specific ingress and egress points in the network? End-to-end performance for the service?
- Are end-to-end service guarantees (SLGs) a requirement (e.g. is end-to-end QoS per traffic class a requirement)? If so, what are the queuing/buffering mechanisms used for modeling the network performance?
- Are network performance reports done by service or by customer or aggregated per network node or per network application or by protocol used?

- If network performance reports are by a customer, are there any internal performance measurements that are outside the customer SLA/SLG reports?
- Are there any billing implications as a result of an SLA/SLG not being met? Do the customers'/end-users' bill payments depend on the network performance measured?
- What are the coloring schemes for the different types of network performance characteristics on the graphical displays?
- Who will be the audience of the network performance reports?
- What are the nonfunctional requirements for the system? What is the acceptable delay from system users requesting an analysis or report to the system producing the results? What load conditions will these response times be under?
- Which network performance data/parameters are to be displayed, and using what format?

With regard to *system performance*, below are some additional questions (to those above) to consider:

- Which systems require performance monitoring?
- Will these performance measurements have an impact on the system?
- What are the overall system performance targets?
- Are trend analysis and growth prediction requirements for the service/system? If so, what growth factors/assumptions does the system need to make or can the system users enter into the system?
- What are the systems parameters to be monitored?
- Will the system performance be measured against any service performance/KPI of the service?

Further details and questions on system performance are covered in Section 8.13.2.

As services and applications using the telecommunication network become more sophisticated, SLGs are what the customers and end users are looking for. This is especially true for mobile services, where applications form an important part of the service. Therefore, ways to measure and monitor application response times are becoming more important. When considering *application performance* monitoring, it is important to decide upon:

- Which applications require performance monitoring?
- What performance parameters are to be measures?
- What is the performance target for each parameter?
- What are the levels at which performance degradation will be noticeable to the end users/ customers and actions are to be taken?
- What are the thresholds for each performance parameter to trigger alarms in the performance management system?
- What is the approach for measuring different applications? Will it require an active approach or is a passive approach more appropriate?

Application performance approaches and example performance measures can be found in Section 8.13.3.

8.13.1 Network Performance

The major system function for network performance management is providing the facility/tool to monitor network performance parameters. As part of the network management functions (as stated in Section 8.12.1.1), the network is polled for network performance data. Over time, trend analysis can be performed. The following functional areas should be considered:

- define network performance parameters;
- collect and store network performance data;
- monitor network performance and threshold levels;
- track network KPI violations;
- define reporting format;
- generate reports and display reports;
- monitor network trends;
- perform trend analysis;
- model network resources;
- simulate network behavior and performance under predefined conditions and scenarios.

The relationship between network performance and service performance is well summarized in Section 3.1.9 of the *SLA Management Handbook* [55].

8.13.1.1 Define Network Parameters

To produce network performance reports, the following definitions should be considered:

- define network elements and their interfaces to monitor;
- define performance parameters and thresholds to monitor (e.g. link utilization, network equipment processor utilization, traffic profile, round-trip delays, packet loss and delay variation/jitter, queue depth and buffering);
- define the network performance characteristics and thresholds to monitor;
- define the network KPI to be monitored;
- define traffic types to monitor;
- define end points – source (traffic ingress points) and destination (traffic egress points).

8.13.1.2 Collect and Store Network Data

Data collection and storage functions include:

- collect network performance data from nominated network nodes/equipment;
  - ○ define collection intervals;
  - ○ define parameters to be collected;
- probe network elements for performance data;
  - ○ define polling intervals;
  - ○ define parameters to be polled;
- validate data to ensure information is complete, genuine and uncorrupted;
- store network performance data (present and historical).

8.13.1.3 Monitor Network Performance and Threshold Levels

The following network performance monitoring functions should be considered:

- monitor network performance parameters and thresholds;
- monitor the QoS levels (where required);
- monitor the impact of engineering work, such as fault clearance;
- detect degradation of network performance thresholds and generate alarms in the network management system where relevant.

8.13.1.4 Track and Predict Network KPI Violation

From the monitoring function above, these network KPI violations are tracked to ensure potential problems in the network are resolved. The service management system should be informed about these SLA violations. The system function should include the prediction of network KPI violations through trend analysis (please see below for trend analysis functions).

8.13.1.5 Define Reporting Format

Formats of the network performance reports are normally graphical. The graphical displays are in the form of graphs or tabular format. These graphs and tables should be made simple for ease of understanding and to minimize misinterpretations. The graphical displays are normally in color for ease of reference and for spotting problems. Typically, RAG reports are used:

- R – red for critical, immediate attention required;
- A – amber for warning, some action is required;
- G – green for OK, no action required.

Criteria/thresholds for each RAG category are to be defined.

8.13.1.6 Generate Reports and Display Reports

To generate network performance reports, the following functions should be considered:

- define frequency and granularity of reports;
- produce predefined reports at the predefined frequency;
- perform analysis as per system user's request;
- produce ad hoc reports as per systems user's request;
- display graphically (or tabular or as specified) the performance of the network in accordance to the predefined performance parameters, network performance characteristics and performance targets and thresholds;
- save and store report (if required).

8.13.1.7 Monitor Network Trends

The following functions should be considered for trend analysis:

- define network parameters for trend analysis;
- define mathematical formula to be applied to perform trend analysis;
- define the graphical format and granularities of trend analysis results;
- define network trend thresholds levels;
- perform trend analysis;
- display graphically the trend analysis results;
- monitor network trend thresholds;
- highlight trends that will exceed thresholds if no action is taken.

8.13.1.8 Model Network Resources

If network modeling is required, then the following system functions should be considered:

- define nodes to be included in the network model;
- define model input data (e.g. the network parameter/data);

- input network modeling techniques and formulas to perform mathematical analysis to quantify network behavior;
- apply network performance modeling technique to model and quantify network behavior;
- input network design rules and map to form a representative model of the real live network.

8.13.1.9 Simulate Network Behaviors and Performance Under Predefined Conditions and Scenarios

Network simulation is useful for validating network design and network upgrade activities. The following system functions will be useful:

- simulate routing changes in the network under failure scenarios;
- simulate network performance due to additional network node;
- simulate the effect of rerouting delays due to route failures;
- predict network performance impact on customers under failure scenarios;
- predict network impact as a result of changes in traffic mix;
- simulate network performance under load;
- perform impact analysis on the introduction of new services and technologies.

8.13.1.10 Example Performance Parameters for PS Network

Below are some examples of performance parameters you might want to consider for packet switch services:

- packet transfer/round trip delay;
- round-trip time;
- transfer delay variation for IP packets;
- packet loss ratio;
- duplicate packets;
- latency/delay variance;
- link utilization;
- session throughput;
- CPU utilization of network equipment;
- link/node failure;
- network equipment availability – measure in terms of MTBF (in hours or years); or MTTR or mean time to failure (MTTF; i.e. up-time measured in hours); or mean downtime (MDT; in hours); or failure rate;
- equipment memory utilization.

8.13.1.11 Example Performance Parameters for CS Network

Below are some examples of performance parameters you might want to consider for CS services:

- call set-up delay (length of time required from session/call set-up request to session/calls being established);
- call clearing delay (length of time required from the call clearing request to the call/session being terminated);
- set-up message transfer time (time required to set up the signaling link after the set-up message was received by the node);
- signaling transfer delay (time taken to transfer a message from one signaling node to another);

- signaling message loss ratio (proportion of signaling messages that are lost);
- echo;
- cross talk;
- circuit utilization;
- BHCA;
- total call completion;
- total call loss;
- average duration of call;
- circuit/port utilization.

8.13.1.12 Additional Performance Parameters for Mobile Access Network

Example performance parameters for a mobile access network may include:

- lost calls percentage per cell group in regions;
- access failures (deny end-user access to network) per cell and per region/sector;
- bit error rate, frame error rate and signal quality estimate per cell;
- signal strength and transmission loss per cell and per region;
- radio blocking (due to congestion) per cell;
- usage and RF loss per cell;
- handoff failures per region;
- authentication delay (time required to authenticate a mobile end-user).

Below are some examples of performance parameters you might want to consider for mobile services:

- maximum bit rate per end user;
- guarantee bit rate per end user;
- maximum packet size;
- packet ordering (are packets delivered in order?);
- traffic handling priority (used to create subclasses within the interactive class);
- reliability (are error packets delivered?);
- session throughput.

Additional challenges for 3G mobile service performances include:

- measuring and monitoring end-to-end QoS/SLA across multiple networks and technologies;
- performing impact analysis on end-users' experiences during the occurrences of IP QoS violations;
- modeling and carrying out performance analysis on IP-based service over 3G networks.

For mobile network performance parameters and guidelines, please refer to the *Wireless Network Performance Handbook* [24] or *UMTS Performance Management: A Practical Guide to KPIs for the UTRAN Environment* [61].

8.13.2 System Performance

As you can see from this chapter, system solutions play an important role for all aspects of service operations. Hence, ensuring that the systems are performing at their optimum efficiency levels is vital to the success of operating the service and gaining customer satisfaction.

To analyze systems performance, performance metrics are to be defined, measured and analyzed. The performance metrics are divided into two main areas: hardware performance and system overall performance. The hardware performance metric measures the hardware utilization and performance levels, whilst the system overall performance metric measures the efficiency of the systems.

When designing the system performance management tool, nonfunctional requirements for the system will be the main drivers and considerations. Each system will probably have a different performance target. All these should be defined at the start of the development/specification of the system performance tool. The following questions should help to capture and further define the requirements for the system performance management tool:

- What are the objectives for system performance? To satisfy internal performance KPI? To satisfy customer SLAs?
- What are the systems to be monitored?
- What are the system parameters to be monitored?
- What are the system performance targets (mean/maximum or minimum threshold values) for each of the systems to be monitored?
- What is the tolerance level on processing time for a certain task/function/request after input has been submitted into the system?
- What is the system availability requirement? Is the system availability requirement on a specific system (or subsystem) or the integrated system as a whole? If the system is a vital part to the service, how will this availability figure affect the overall service availability figure/requirement?
- What are the reliability requirements to achieve the system/service availability figure?
- Are there any performance measurements required at the interfaces of the systems?

The main system functions to measure system performance are similar to those of network performance mentioned above. To minimize duplication, I have cross-referenced the relevant sections for detailed system functions against the system functions that are the same as below.

- define system parameters to be measures;
- collect and store system performance data (please see Section 8.13.1.2);
- monitor system performance;
- define performance reporting format (please see Section 8.13.1.5);
- generate performance reports and display reports (please see Section 8.13.1.6);
- monitor trend analysis (if required);
- perform trend analysis;
- simulate system behavior and performance under predefined conditions and scenarios (if required);
- store reports produced (if required).

8.13.2.1 Define System Parameters

In order to find out how the systems are performing, the system performance parameters to be measures are to be defined. The following definitions should be considered:

- Define performance parameters and thresholds to be monitored (e.g. CPU utilization, network link utilization, memory utilization). Please see later sections for further parameter examples.
- Define system performance characteristics and thresholds to be monitored.

8.13.2.2 Monitor System Performance

To monitor the system performance, the following monitoring functions may be required:

- monitor system performance parameters and thresholds;
- monitoring the QoS provided to major customers (if required);
- monitoring the impact after system or software upgrades;
- detect degradation of system performance parameters and thresholds.

8.13.2.3 Monitor System Trends (If Required)

System trends normally apply to system hardware and system load. Other performance parameters are likely to be system or service specific. It will also be useful to map system events or upgrades to the trend analysis to spot potential system performance problems or to see whether certain system problems have been resolved. Please see Section 8.13.1.7 for functions to be considered for trend analysis.

8.13.2.4 Simulate System Behaviors and Performance Under Predefined Conditions and Scenarios (If Required)

This function is required for complex systems. This will help to define the hardware upgrade requirements. System simulation may also be useful for preparation work for system software upgrades where there is a significant change in the way in which the new software is going to use the system resources available.

8.13.2.5 System Hardware Performance Parameters Examples

The following are some system hardware performance parameters you should consider measuring

- CPU utilization of system equipment;
- system memory usage and utilization;
- disk space usage and utilization levels;
- system interfaces utilization.

8.13.2.6 System Overall Performance Parameters Examples

Apart from the hardware performance of the system, one may want to consider monitoring the performance of the various transactions taking place in the system. For example, in an order management system, one may monitor the time taken to accept an order or process a cancelation order and so on. System performance measurements may be for part of the KPI of the service, but should definitely be part of the nonfunctional requirements of the system.

System performance indicators may include:

- average time required to perform a certain transaction/function/system request or system response (or time in terms of percentage, e.g. transaction time is less than 3 s, 90% of all requests);
- number of parallel transactions in a typical minute;
- average elapsed time to perform a certain transaction;
- system throughput on certain transactions/work strings;
- system load (peak and average);
- system I/O utilization;
- system availability over a defined period (monthly and/or yearly);

- error rate (failures per day/per hour/per year) for transactions;
- system availability;
- MTBF/MTTF and MTTR and time to repair figures of the system;
- MDT of the system;
- number of occurrences where data not available to system users'/customers' requests;
- response time for predefined functions with system load;
- time required to produce required reports (average and maximum time) together with system load at the time of measurement;
- memory usage/utilization;
- number of transactions per second/per minute;
- thread/job queue length;
- number of rows of data being processed per second;
- number of requests per second.

8.13.3 Application Performance

Before one can decide on the solution to monitor the performance of applications, you need to decide on the approach as to how the performance for each of the applications is to be measured. The following approach can be considered:

- setting up network probes at various parts of the network to monitor the performance parameters (e.g. maximum, average and minimum response times) of application traffic at various points of the network;
- carrying out active monitoring of application transactions by placing test transactions that imitate the real transactions on the network at regular intervals to measure application performance parameters;
- using performance probes at user/client application to capture application performance data.

The approach to use depends on the application concerned, the performance data required and the scale of operation. Analysis and considerations of these approaches can be found in *Managing Service Level Quality Across Wireless and Fixed Networks* [32].

Examples of application performance measures are listed below:

- application response time;
- network response time;
- application up-time and downtime;
- network up-time and downtime/network availability;
- server/Web service response time;
- number of end-user complaints.

Once the approach for measuring the application performance has been defined, system functions to collect application performance data and to perform the application performance analysis that are similar to that of the system performance can be applied. Please refer to Section 8.12.2 for details on system functions.

8.14 Capacity Management, Traffic Management and Network Planning

The system functionalities for capacity management are very similar to those of performance management. However, capacity management functions focus on the analysis of the performance measurements and interpretation of data such that the relevant operations personnel can proactively forecast capacity deficiencies and schedule sufficient capacity enhancements to meet the SLAs or SLGs that have been contracted with customers/end users before performance degradation occurs. This concept is relevant to both network and systems capacity within a service. To improve efficiency in gathering and storing data, the data for capacity management reporting should be included as part of the network performance monitoring parameters.

Capacity management tools also provide information to enable capacity planners to optimize the network utilization and maximize the usage of network asset.

With the above in mind, what are the capacity management requirements on the system? What questions should we be asking?

8.14.1 Network Capacity Management Systems Requirements Questions

- Is a network modeling tool required? If so, what network elements require modeling?
- What network routes require simulation?
- Which network traffic modeling methodology/traffic analysis methodology is used to analyze the capacity requirements?
- What scenarios (e.g. failures scenarios or 'what if' situations) are to be considered for capacity management purposes?
- What are the utilization thresholds for each network element/interface before performance degradation occurs?
- What are the utilization thresholds for each network link before performance degradation?
- What are the tolerance levels for each network element/interface before performance degradation becomes noticeable by customers/end users?
- What are the performance characteristics of the network link?
- What combination of these characteristics signifies potential network performance degradation?
- What are the planning guidelines/planning rules for the service?

8.14.2 Network Capacity Management System Functions

Considering all the above, below are the main system functional areas required for network capacity management:

- define resources that require management (e.g. network elements, IP addresses, number ranges);
- define capacity threshold reports for all resource to be capacity managed;
- monitor utilization and load of defined network resource;
- produce network capacity reports;
- perform trend analysis;
- input network capacity forecast data;
- forecast potential capacity shortfall;
- define reporting format;
- define reporting intervals (for ease of comparison, the reporting format and reporting intervals should be consistent between different traffic types).

8.14.2.1 Define Network/Resource Capacity Reports

In order to produce network capacity reports, the following are to be defined.

- define network/resource capacity and utilization parameters and thresholds to be monitored;
- define reporting frequency and granularity of utilization reports;

- define all the threshold levels of the RAG reports, including critical capacity thresholds, before performance degradation becomes noticeable by customers/end users, where immediate attention is required;
- define underutilization levels for each network element or network link or resource.

8.14.2.2 Monitor Network/Resource Utilization and Network Load

To predict the capacity demand in the network, one needs to know the current utilization and load levels. Therefore, the following system functions are necessary:

- define the network links or network elements or resource to be monitored;
- define the utilization threshold levels and load parameters for each traffic type for each network link/interface or network element to be monitored;
- define data collection intervals;
- enter the formula to calculate utilization and load levels;
- collect the parameter values from the relevant network elements;
- validate data to ensure information is complete, genuine and uncorrupted;
- calculate utilization and load levels for all the defined network links/elements;
- highlight areas that are underutilized or above threshold levels.

8.14.2.3 Produce Network Capacity Threshold Reports

Using the utilization data collected, reports are produced to highlight areas where the network has a high utilization and areas where the network is underutilized. The following systems functions should be considered:

- generate utilization reports;
- display graphically (using RAG reports) the utilization of the network elements or network links in accordance with the predefined utilization parameters and thresholds;
- identify potential congestions/bottlenecks in the network;
- identify areas where network is underutilized.

8.14.2.4 Perform Trend Analysis

Simulating growth trends will be helpful to plan network capacity. The following functions may be useful:

- define growth factors for network link or node or network element;
- define customer traffic profiles for the service;
- simulate growth trends in accordance with predefined growth factors.

8.14.2.5 Forecast Potential Capacity Shortfall

For network planners, it will be useful to forecast potential capacity shortfalls. The following tasks/functions should be considered:

- collect provisioning network data;
- input marketing forecast data;
- input orders in the 'pipeline';

- collect input from trend analysis;
- perform analysis to predict when the capacity shortfall is going to arise, taking into account all the above four factors;
- dimension capacity enhancement;
- schedule enhancements;
- simulate network performance with and without enhancements by applying growth factors on the network.

8.14.3 Capacity Management for Mobile Access Network

The main difference between the capacity management function for fixed and mobile networks is the capacity management of the RAN and functions associated with RF planning. In most 3G environments, the most expensive asset is the RF allocation in the access network; hence, careful planning and utilization of such an asset is very important. Additional system functions for capacity management for mobile access networks include:

- perform frequency planning for each cell site area;
- estimate and establish the number of radio masks that are to be added or removed from each cell site area;
- monitor the growth rate and actual additional capacity made available;
- producing the following reports may be of use:

  - radio frequencies used at each cell site;
  - Erlangs usage/carried per cell site;
  - total number of channels available and used per cell site;
  - number of activated radio masks/carriers at each cell site;
  - under radio exhaustion scenario, the percentage of blockage compared with target set as part of the design rules;
  - channel exhaustion at each cell site;
  - offload factor used for shedding traffic to another cell site;
  - the amount of acquired Erlangs from neighboring cell sites;
  - total number of channels required;
  - total number of radio carriers required.

For further details on mobile capacity management and reports, please see Chapter 10 of the *Wireless Network Performance Handbook* [24].

Within the 3G core network, most of the network capacity management system functions listed above do apply. An additional/new capacity management model may be required for IMS, as it handles a mixture of voice calls and data sessions, as well as different content and application types.

8.14.4 Network Traffic Management

Network traffic management and capacity management are closely linked. In order to achieve the optimum network utilization level and the required network performance requirements, the network capacity is to be provided at the right place in the network and the network traffic is to be sent to where network capacity is available. Therefore, the system needs to make the network traffic data visible to the traffic engineers so that they can make the right routing decisions for the network traffic.

Lack of traffic management may lead to:

- loss of revenue due to network congestion;
- degradation of service, resulting in breach of customers'/end-users' SLAs;
- customers'/end-users' bad perception of service quality;
- unnecessary capital spend on additional network capacity.

When designing the systems function to support this area, one may consider the following:

- Do you measure the traffic flow by service, by customer or aggregated level in the network?
- What are the parameters that are meaningful measures of traffic flow?
- Where will the data come from?
- Is real-time statistics collection and analysis required? If so, what are the data collection intervals and analysis window?
- What is the frequency of measurements/data to be collected for nonreal-time reports?
- What is the frequency of nonreal-time reports?
- Is ad hoc report generation a requirement?
- Do you measure traffic flow by traffic type?
- If QoS is applied in the network or service, do you measure the traffic flow per QoS category?
- Does the system need to redirect traffic flow dynamically to a different part of the network if congestion is occurring in one area? If so, what is the mechanism for doing that?
- Do you need a network traffic flow model? If so, which one (e.g. peer-to-peer, client–server or hybrid of the two) is most appropriate for the service/network concerned?
- Do you measure the traffic by customer/end user or per application or between network elements?
- How long do you keep the data for?

The system functions to support traffic management are similar to that of capacity management, but there is some difference in the details.

8.14.4.1 Define Traffic Parameters to Be Measured

The following are to be defined for traffic management reports:

- define traffic types to be measured;
- define the traffic flow parameters for each traffic type to be measured;
- define frequency of measurements and data collection intervals;
- define data sources for traffic measurements;
- define parameter thresholds for potential traffic problems;
- define parameter thresholds for traffic congestion.

8.14.4.2 Collect and Store Traffic Data

To provide traffic management reports, traffic data is to be collected from the predefined points/ network elements/data sources. The following system functions are required:

- collect the traffic flow, traffic rate, burst size and other traffic engineering parameters;
- validate data to ensure information is complete, genuine and uncorrupted;
- define the data store period;
- store current and historical data.

8.14.4.3 Perform Traffic Analysis

The following traffic analysis will be helpful:

- compare parameter values with thresholds set;
- identify where thresholds have been exceeded;
- identify congestion/hot spots in the network;
- identify source and destination/sink of traffic;
- identify traffic patterns, distribution and trends;
- identify worst-case scenario and benchmark normal traffic scenario.

8.14.4.4 Create Traffic Model

Traffic modeling is useful to traffic engineers in order to help them to understand what is actually happening in the network and ascertain potential traffic growth. The following system functions can be considered:

- apply mathematical formula/model to convert source data to required traffic information;
- model the traffic flow of the network and volume of traffic ingress and egress from the predefined points of the network;
- apply traffic management theory/model to predict traffic growth based on historic data;
- revise routing of traffic at a certain network node and simulate network behavior;
- convert result to visual display (e.g. graphical, geographic display on a map).

8.14.4.5 Activate Traffic Mechanism

Where relevant, the system should have the capability to activate/initiate predefined traffic control procedures/mechanisms automatically when certain criteria/traffic levels have been exceeded. This should help to ease congestions in the network.

8.14.4.6 Define and Produce Traffic Management Reports

The following functions are required to produce traffic management reports:

- define reporting period (e.g. hourly daily or weekly or monthly);
- define reporting format;
- define granularity of the reports;
- define targets for each traffic parameter;
- define RAG indicators for each parameter;
- produce RAG reports;
- graphically map out the traffic flow and distribution of network traffic.

Details on traffic management measurements can be found in *ITU E.490.1 Traffic Engineering – Measurement and Recording of Traffic: Overview Recommendations on Traffic Engineering* [23].

8.14.5 Network Planning Tools

The main functions for network planning tools are similar to those of the network performance tool mentioned in Section 8.13.1.9. However, network planning functions should also include:

- importing various sources of network-related data (market trend and strategy, network capacity data and sale forecast) to form network forecast and network plans;
- network behavior simulation based on 'what if' scenarios.

For a 3G service, network planning tools will need to take account of:

- RAN planning;
- network planning and dimensioning for both CS and PS networks in combination with the RAN planning;
- different simulation techniques to cater for the 3G network.

Further details on 3G network planning can be found in *Fundamentals of Cellular Network Planning and Optimisation* [49].

8.14.6 System Capacity

To manage system capacity, the questions to consider include:

- What are the systems capacity variables?
- What are the utilization thresholds for each system parameter before performance degradation occurs?
- What is the system behavior if system capacity is running low?
- What are the symptoms/warnings when the system capacity is low?
- What system parameters will need to be monitored for system capacity management?

System capacity can be a contributing factor to system performance. Apart from the system hardware capacity, other system capacity items may include:

- number of transactions/requests/orders/faults, etc. the system can process or handle within a defined timeframe;
- number of historical records to be held and size of database;
- memory usage;
- disk space;
- processor activity and utilization levels.

From a system capacity monitoring and analysis point of view, the system functions to support this are exactly the same as those of network capacity management. The main differences are the definition of the systems to be monitored, the parameters to be measured and the thresholds to be set. It is not unusual to use the same tool, but probably on separate systems, to monitor both the network and system capacities. Some of the hardware parameters to be measured (e.g. CPU utilization, memory utilization) are the same. Therefore, the high-level system functional areas required for system capacity management are:

- define system capacity threshold reports;
- monitor system capacity, utilization and load;
- produce system capacity reports;
- perform trend analysis;
- forecast potential capacity shortfall.

For detailed system functions, please refer to Sections 8.14.1.1–8.14.1.5.

8.14.6.1 Example System Capacity Parameters

Below are some example parameters for measuring monitoring system capacity:

- CPU utilization;
- memory usage and memory utilization;
- I/O utilization;
- number of transactions/requests/orders/faults, etc. the system can process or handle within a defined timeframe;
- number of historical records to be held and size of database;
- disk space;
- processor activity and processor utilization levels;
- system processor queue length.

Together with the system performance parameters defined in Section 8.13.2.6 and the items to be monitored of a system in Section 8.16.1.6, you should be able to have a very good view of the system capacity status and be in a position to assess whether additional system capacity is required.

8.15 Reporting

Reporting plays an important role in the management of the service. Without service performance reports (KPI reports) and the customer reports on SLA status, it would be very difficult to manage the performance of the service. Using service performance reports to monitor service performance is vital to give customer satisfaction and to provide a good service. Most reports for the service are linked to the state of health of the service. Therefore, most reporting requirements are specific to the service concerned. However, the general questions to be asked when designing a reporting system and defining the system functions to produce reports are very similar.

Despite the importance of reporting, it is often left till last when defining and designing the service. Most people think that reports just appear from the system. Well, they do so only if the data is available in the right format and held in the right place in the database. The database needs to be sized and structured in such a way that all the data for the reports can be combined sensibly together without too much overhead. Reporting requirements often dictate the database structure for the system. Therefore, without considering the reporting requirements when designing the system/service, there is no guarantee that the data is going to be available at the right place and that the reports can be produced. Below are some questions to help capture the reporting requirements in general. Nonfunctional requirements, especially relating to size of data and response times for producing reports, are very important.

- What is the purpose of each report?
- Who are the audience for the reports?
- Who is responsible for the report generation?
- What will happen if the reports are not generated?
- How will the reports be used?
- What reporting items are required?
- Where will the data be coming from?
- What is the format and size of input data?
- How often will the data be collected?
- When should the reports be produced?
- What are the formats (graphical or tabular) of the reports?

- What types of graphical display/interface are required?
- What is the granularity of the reports?
- Is analysis of the data required? If so, what type of analysis and form of data manipulation are required?
- What are the analysis and reporting periods (daily/monthly/yearly) for these reports?
- What is the reporting medium (electronic/paper)?
- How will the report audience access the reports (log on to reporting system or report posted on a shared access area)?
- What is the report distribution mechanism (e-mail, sent out on CD)? Who controls the distribution of the reports?
- Will the report contain sensitive information? If so, what is the most appropriate security mechanism to be applied to the reports?
- How long does the data need to be stored for? How much historic data is required?
- Is ad hoc reporting a requirement?
- If ad hoc reporting is a requirement, what data is to be made available for system users to generate these reports? How frequent will these ad hoc reports be generated?
- What is the maximum acceptable time it takes to generate ad hoc reports and under what load condition?
- What are the system response times (and under what load conditions) to the request of producing predefined reports?

8.15.1 Reporting System Functions

The main system functions for reporting are:

- define reporting parameters, data format and collection intervals;
- collect and store data;
- perform required analysis;
- generate report in the defined reporting format and medium;
- store reports;
- distribute reports or make them available to audience.

8.15.1.1 Define Reporting Parameters, Data Format and Collection Intervals

System functions for report definition may include:

- define data/parameters to be measured;
- define format of data;
- define frequency of measurements and data collection intervals;
- define data sources for data to be collected.

8.15.1.2 Collect and Store Data

System functions for collecting and storing data include:

- collect the data from the relevant network elements or systems;
- validate data to ensure information is complete, genuine and uncorrupted;
- define the data store period;
- store current and historical data.

8.15.1.3 Perform Required Analysis

If analysis of the data is required, then the following functions should be considered:

- define data/parameters for analysis;
- define mathematical formula to be applied to perform the analysis;
- define the graphical format and granularities of analysis results;
- define thresholds levels of result (if any);
- perform analysis;
- display graphically the analysis of results;
- highlight items that exceed thresholds.

8.15.1.4 Generate Report in the Defined Reporting Format and Medium

To generate the reports, the following functions should be considered:

- define frequency and granularity of reports;
- produce predefined reports at the predefined frequency;
- produce ad hoc reports as per systems users' requests;
- display graphically (or tabular or as specified) the reports in accordance with the predefined formats and reporting intervals/times.

8.15.1.5 Store Reports

To store the reports, the following may need to be defined:

- naming convention of reports;
- storing and archive period;
- storage area for reports;
- format of stored reports;
- media on which the reports are stored;
- the mechanism to retrieve the stored/archived reports.

8.15.1.6 Distribute Reports or Make Available to Audience of Reports

If the reports are to be distributed to the report audience, the mechanism of distribution is to be defined and acted upon by the system accordingly. If the reports are stored in a shared or predefined file location, then appropriate security and authorization are to be put in place to ensure the correct reports are accessed by the right people.

8.15.2 Report System Function Summary

System functions for reporting are summarized in Figure 8.13.

8.15.3 Example Management Reports (Customer Reporting, SLA and KPI Measurements)

8.15.3.1 SLA and Customer Reports

Depending on the service, the following examples of customer/customer SLA reports may be useful:

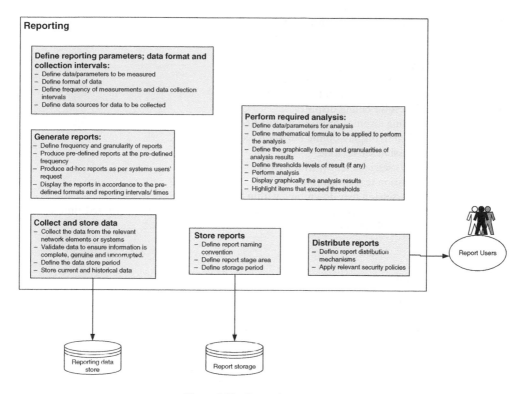

Figure 8.13   Reporting summary.

- number of end users on the service;
- number of end-user orders per week/per month;
- average service provisioning lead time;
- percentage of orders provisioned within SLA;
- number of faults per month;
- average response time for a fault reported by a customer;
- percentage of faults responded within customer SLA;
- percentage of faults resolved within customer SLA;
- average network change request lead times;
- percentage of network change request completed within SLA;
- average service change request lead times;
- percentage of service change requests completed within SLA;
- average response times for end-user authentication;
- percentage of response times within customer SLA;
- number of failed authentication attempts;
- number of successful authentication attempts;
- average time taken to display online customer reports upon request;
- percentage of time taken to deliver customer reports on time;
- average time taken to download online customer reports upon request;
- average time taken to update customer fault reports upon request;
- percentage of time taken to update customer fault reports within SLA;

- average time taken to update customer order status upon request;
- percentage of time taken to update customer order status within SLA;
- service availability (monthly/yearly);
- billing accuracy as a percentage;
- percentage of bills delivered on due date;
- number of customer billing disputes per month;
- average time to resolve billing disputes;
- percentage of billing disputes resolved within 30 days;
- average time taken to display online bill upon request;
- average call or session set-up time;
- percentage of calls/sessions set up within SLA;
- average call or session tear down time;
- percentage of call/session tear down time within SLA;
- daily network latency measurements taken between defined points at defined time of day charted over a monthly period;
- packet loss.

8.15.3.2 KPI Measurements

KPI reports may include:

- the revenue generated from the service in a year;
- the number of new customers or end users per month;
- the number of customers/end users retained per month;
- the number of customers/end users leaving the service per month;
- the number of customers/end users returning to use the service per month;
- the number of customers/end users provisioned per month;
- the number of customers/end users provisioned late per month;
- the percentage of customers/end users provisioned late per month;
- the average, longest and shortest times to provision the service for a customer/end user (from contract signature to service installation complete) per month;
- the percentage of service provisioning within the customer SLA;
- the number of faults per fault category per month;
- the average, longest and shortest times to fix a customer/end-user reported fault (per category of fault) in a calendar month;
- the percentage of faults being fixed within the customer SLA – different category of fault might have a different percentage;
- the number of customer/end-user billing enquiries per month;
- the percentage of bill enquiries that are dealt with satisfactorily within the customer/end-user SLA;
- the number of technical support calls received per day/per week;
- nature of the technical enquiries (break down by percentage per category);
- the average time to resolve technical enquiries;
- the service credit payout per month as a result of breaching agreed SLAs;
- the number of bills and amount outstanding more than 90 days;
- the average time required to terminate/deprovision a customer/end user on the network after service cancelation/termination requests have been received within a calendar month;
- the average, longest and shortest times it takes before a customer service agent answers a call when a customer/end user rings the service help line;
- network and system performance reports are also candidates for KPI, depending on the service.

8.15.3.3 Performance Reports

Performance reports are mainly divided into the following areas: network performance, system performance, operational performance, financial performance and service performance. Performance reports should include any known risks to the service/network/systems/operational areas. For some example reports on network and system performance, please refer to Section 8.13.

For operational performance, the service KPI and customer SLA reports in the previous section provide a good indication on the operational performance for the service. Additional service performance measurement may include customers'/end-users' satisfaction with the service.

Financial performance for a service is mainly concerned with the revenue generated by the service. A list of revenue reports can be found in Section 8.10.1.

8.16 System Support and Management

With more services moving towards centralized service profile management and application-based solutions, managing system availability is just as important (if not more important in some services) as managing the network. The availability of the systems will have a direct impact on the availability of the service and, hence, revenue generation opportunities. The effect on customer satisfaction and service degradation can also be very noticeable. Therefore, it is important to ensure that the systems are available and running efficiently and that they are monitored for any potential failures.

The main areas for systems support and management include:

• system monitoring and event management;
• system fault management;
• system disaster recovery and configuration management;
• system administration;
• system maintenance;
• system performance and capacity management;
• system security.

8.16.1 System Monitoring and Event Management

The system monitoring functions are very similar to that of the network management functions stated in Section 8.12. Most systems also use SNMP to communicate status information to their management system. The system monitoring concept is much the same as the network. The major difference, apart from the devices and parameters to be monitored, is the functionality for alarm correlation, as problem areas are more apparent and notifications from the problem system provide good indications as to the source of the failure.

The principal system functional areas for system monitoring include:

• define systems to be monitored;
• monitor system status and events;
• filter and process system notifications and events;
• alarm categorization and troubleshooting;
• perform troubleshooting on the system.

8.16.1.1 Define Systems to be Monitored

Before any system can be monitored, one needs to:

• define the systems to be monitored (which involves creating an object to be monitored in the system monitoring system);

- define system interfaces and parameters to be polled/monitored;
- define polling intervals;
- define parameter thresholds (if event is to be generated when thresholds are exceeded).

8.16.1.2 Monitor and Logging System Status

Monitoring system status involves proactively gathering system status from the defined systems. This involves polling all the systems to obtain status information. The major functions for monitoring the systems status are:

- poll the systems for status information;
- collect polling data from the systems with date and time stamp;
- validate data to ensure information is complete, genuine and uncorrupted;
- store system polling data.

8.16.1.3 Monitor and Logging System Notification/Events

Systems send out notifications to indicate potential problems that may occur. These notifications can be notification of potential problems, of failure of system hardware or interface, and that certain preset thresholds or conditions have been reached or exceeded.

The major functions for monitoring the system events are:

- collect system notification (i.e. SNMP traps);
- validate data to ensure information is complete, genuine and uncorrupted;
- log system events.

8.16.1.4 Filter and Process System Notifications/Traps and Events

The following functions are carried out as part of filter and process system notifications:

- define system event filters and filter configuration;
- apply filters to system traps (the filters can suppress notification as faulty systems can generate too many notifications);
  - o user filters may be applied, as each system user may only want to view systems that they are responsible for;
- translate system traps to network events;
  - o translate SNMP traps (using the definition in the MIB) to a more meaningful event type;
- generate system event if characteristics exceed predefined thresholds/conditions;
- forward system events to defined management system or person/system user.

8.16.1.5 Alarm Categorization and Trouble Shooting

The system needs to provide functionalities to help the system users (normally system management personnel) to diagnose the system faults. Alarm categorization helps to prioritize system events. When designing the system monitoring system, one may consider the following functionalities:

- categorize the severity of events into alarms and alerts;
- define the format of graphical display of alarms;
  - o most 'critical' alarms are labeled red such that the alarms can be spotted easily; others are up to the users to define;

- define alarm groupings;
- group alarms according to groupings or pair up events to indicate source of fault;
- monitor repeated faults and recurring problems;
- display of alarms in a defined grouping and categorization;
- provide GUIs for the system user to access/drill down to the desired system components for additional historical information for troubleshooting and fault isolation;
- view system logs/historical events and details of system components.

Alarm severity categories can be defined as 'critical' (i.e. require immediate attention), 'major' (i.e. actions required as soon as possible), 'minor' (i.e. potential service degradation, monitoring required) or 'warning' (i.e. potential problem may occur, investigation required).

8.16.1.6 System Monitoring Parameters Examples

System parameters and items to be monitored may include:

- memory usage, including RAM, virtual memory usage;
- disk space availability;
- disk performance, including average disk queue length, disk read and write latency, disk read and write per second;
- processor activity and utilization level for each processor thread (%Processor Time);
- I/O utilization;
- amount of network data transmitted and received;
- network interfaces, including the amount of data incoming and outgoing, network link status, packet received errors and packet send errors;
- link queue, which holds all the messages to be transmitted to another system – if the link queue is getting large, then it is an indication that the processes that send out the data are faulty or there is a network link problem;
- system queue, which holds messages that are awaiting processing – if messages remain in the system queue for too long, then this indicates that there is a problem with the system messaging processes;
- system application operations;
- security breaches;
- abnormal log-ons (e.g. out of hours);
- log-in failure attempts;
- event logs;
- size of event logs;
- system transaction logs;
- size of transaction logs;
- scheduled tasks to be performed;

8.16.2 System Fault Management

System fault management is another area that most people forget about. Since we have established the importance of maintaining availability, providing the ability to log and manage system faults is an important part of the process. System functions for system fault management are very similar to those of network fault management, though nearly all system faults are raised by internal system operational personnel.

The main system fault management functions include:

- fault tickets generation, validation and logging;
- fault diagnostic and fault status update;
- work flow and jeopardy management.

8.16.2.1 Fault Ticket Generation, Validation and Logging

Before a fault can be logged onto the system, validation needs to take place. The following functions should be considered:

- Enable fault ticket input from GUI for system users.
  - Fields for each fault report are to be defined as part of the system definition.
  - If systems are monitored remotely by system operations that are not based in the system operational centers, then external interfaces (Web or others) for the fault management system may be required. This could be over the Internet or intranet.
- Validate that the fault tickets are coming from expected sources.
- Check that there is no duplicate fault ticket (both within the file and within the system).
- Perform simple field validation to ensure that all the required fields are filled with data of the correct format.
- Generate fault ticket reference.
- Log fault tickets.
- Assign fault severity.

  - This is going to be an assessment of the fault by the system operational personnel.

8.16.2.2 Fault Diagnosis and Fault Status Updates

Fault diagnosis and fixing faults are normally performed by human actions. The system should, where possible, perform an initial diagnostic. For example, if the fault is affected by planned outage, or by any known faults or any faults have been reported to third-party support agencies, then the status of the fault ticket should be updated with the expected resolution time. From a system point of view, the following functions should be considered to assist the human actions:

- perform fault correlation where possible;
- identify potential related known system faults;
- log fault diagnostic and resolution;
- log actions performed on the fault tickets;
- update fault status.

Fault status could be:

- open;
- await third-party resolving agencies feedback;
- ongoing investigation;
- closed.

All the status updates should be date and time stamped.

If the system faults are related to a fault raised by the customers/end users, then the systems need to update the status of the fault ticket that is in the fault management system. Cross-references and update automation will be very useful.

If it has been identified that a third-party resolving agency is required to resolve the fault, then the system should the raise fault ticket accordingly. All these faults should be logged and the SLA measured. The following system functions may be required to support raising faults to third-party resolving agencies:

- create fault tickets with the third-party resolving agencies with inter-fault ticket references;
- obtain fault updated from the third-party resolving agencies;
- update internal fault tickets with third-party resolving agencies' updates;
- track third-party resolving agencies' SLAs and alert system users of open fault tickets that are about to breach the SLA.

The above assumes that the third-party resolving agencies have an automated fault logging system.

8.16.2.3 Workflow and Jeopardy Management

As part of the fault management function, the system needs to ensure that the faults are resolved within the internal or customer/end-user SLA and that all tasks are performed by the correct fault resolution agency. The following functions should be performed:

- monitor fault status;
- ensure tasks are performed by the correct resolving agency;
- alert system users for fault tickets that are about to breach a given SLA.

8.16.3 System Disaster Recovery and Configuration Management

System data, configuration and software version must be carefully controlled with regular backups and operational configuration files logged. Backup files are used for system recovery, especially after system failures. The system management system should have the functionalities to enable system upgrades, applied software patches and restarting the system after failures.

8.16.3.1 System Resilience and Redundancy

Within a system, a certain amount of resilience should exist. For example, the system should contain dual processors and dual I/O ports and so on. System redundancy should also be a major consideration when designing a service. A system redundancy strategy can be: having dual systems, running in parallel, mirroring each other's activities or switching to the backup system when the main one fails. Data between the main and backup systems must be synchronized regularly to avoid loss of data. Ideally, the geographic location for the backup/redundant systems should be separate and accessed by different network connections. If the primary site has a disaster (e.g. major power failure), then the backup system will still be operational. This is, of course, dependent on the available budget and risks involved.

8.16.3.2 System Configuration Management

System configuration is very important to ensure the stable operational environment is not in jeopardy and that any changes to the configuration are known and logged. From the system point of view, system configuration management tools can be used. The content of the system config-uration tool can be used in the event of disaster recovery or after system hardware upgrades. Functionalities of the tool may include (but are not limited to):

- logging the existing software version of each of the systems in use;
- storing a copy of the software for each of the systems;
- logging the changes/software upgrades for each of the systems in use;
- backing up the existing system configurations for each of the systems in use.

8.16.4 System Administration

System administration items may include:

- system directory and file management, including the management of shared folders and access permission, audit file system access, evaluate audit policy and event logs;
- database administration;
- transactions logging – for performance and security reasons;
- managing system software licenses.

8.16.5 System Maintenance

System maintenance tasks could include:

- file and data backup and archiving;
- application backup and recovery;
- system shutdown and recovery;
- data recovery after system failure;
- disk management, including the management of disk quotas, performing disk partitioning and defragmenting volumes and partitions;
- manage backup storage media, including access security to backup materials;
- other maintenance tasks as recommended by the manufacturer.

Please refer to the system suppliers' recommendations for system maintenance tasks relating to the system being managed.

8.16.6 System Performance and Capacity Management

To ensure the system is in a healthy state, system performance monitoring and capacity manage-ment are essential. These topics were covered in Sections 8.13.2 and 8.14.5.

8.16.7 System Security

System security is a topic in its own right and I do not intend to cover this in detail here. However, below are some topics that one needs to consider when designing the service:

- systems users' administration and access control – this includes system users'/user groups' account creation, system users authentication (and method of authentication), defining system users' privilege levels and permitting of views and setting users' privilege levels;

- virus protection;
- unauthorized entry into the system (e.g. hacking);
- data obtained from the system without authorization (e.g. spoofing);
- prevention of denial-of-service attacks from unauthorized system users;
- management of remote access to the systems;
- physical security of system equipment;
- system users' malicious attempts, including:

  o conducting denial-of-service attacks on the system;
  o attempting to gain access to another system;
  o attempting to gain access to data that is not within the users' privilege level;
  o posing as another user to gain access to data or conduct malicious attacks.

Further details on system and data security can be fount in *IT Governance: A Manager's Guide to Data Security and BS 7799/ISO 17799* [27], the *Computer Security Handbook* [28], *ISO/IEC 27001:2005. Specification for Information Security Management* [10] and *ISO/IEC 27002:2005: Information Technology – Security Technique – Code of Practice for Information Security Management* [29].

9

Operational Support Processes

Operational support processes define the sequence of events and tasks to be performed to fulfill certain operational or service requirements. Together with the detailed work instructions, they define the processes, tasks and steps to be carried out by operational personnel to perform certain operational functions.

These processes are normally end to end, where the inputs are either from business or external events (e.g. a customer placing an order to order management or network events from network equipment) or from another operational processes (e.g. the fault management process will be an input to the network management process for network fault resolution).

Operational processes are defined in conjunction with the design of operational and business support systems to ensure completeness and that the systems are fit for purpose. If the operational tasks are supported by system functions but require manual input or human decisions, then detailed work instructions will include step-by-step guidance in the use of the systems to perform these tasks. Therefore, both the system designs and process designs go hand in hand and it is virtually impossible to define one without the other, unless the task is fully automated or totally manual.

The operational support processes described in this chapter are split in accordance with functional areas to support the service. These functional areas do not directly map onto the system development areas, but are closely related and are not designed to dictate to the operational organization. These are only logical splits of different functional areas and should be adapted to the operational environment for your needs. The eight operational support processes and areas include:

- sales engagement;
- customer services;
- service and network provisioning;
- service management;
- network management and maintenance;
- system support and maintenance;
- network capacity and traffic management and network planning;
- revenue assurance.

Figure 6.5 in Chapter 6 shows the interrelationships between these processes.

Successful Service Design for Telecommunications   Sauming Pang
© 2009 John Wiley & Sons, Ltd

The development of these operational support processes is as important as designing networks and systems. At the end of the day, it will be the operational teams who will be operating the service and looking after the customers/end users. They need guidelines and processes in order to perform their tasks properly. Different services will have different operational support process requirements. This chapter covers the basic operational processes required to support a service. Readers should change and adapt these processes in accordance with the service and their operational needs. Example operational processes are used to help readers to understand the steps to go through for various operational tasks/functions.

The operational processes categorization is based on their functional areas. One might think that the processes described in this chapter have already been defined in the eTOM [31]. This is true to an extent; however, this chapter concentrates on what processes require designing to operate a service and the conceptual steps within those processes, rather than defining and stating the process itself. However, process examples are included to help readers to understand the tasks involved for certain operations. It illustrates how end-to-end processes can work together across multiple operational and system functional areas to fulfill service requirements. This chapter also highlights the process differences between end user services, which are relatively simple, and the more complex solutions provided to customers. A detailed mapping between the processes described in this chapter and eTOM can be found in Section 9.9.

9.1 Sales Engagement Processes

Sales engagement processes are designed to support the sales teams in identifying sales leads and transforming these leads to orders. Marketing and sales literature (e.g. press releases and a description of the service placed on the company's website) should be produced for the service launch. Through training and the sales literature, the sales teams should understand the service and be able to explain all the service features, the different service options, benefits of all the service options and pricing schemes available and so on to the customers/end users. Marketing promotions and terms and conditions of these promotions should also be part of the training of the sales team.

Other processes, activities and items to be produced to support the sales engagement process include:

- credit checking process – only customers/end users with a good credit rating should receive service, otherwise bad debts will result;
- produce training plans, training materials and provide training for the sales teams and technical sales personnel;
- standard customer presentation for the service;
- order forms for the service to capture all the required details from the customers/end users for the orders (e.g. name and address, service features/options, price plan, payment method, required by date);
- ways to identify up-selling opportunities to customers/end users;
- service launch events and presentations.

Complex customer service and network solutions will require more technical resources. The customers that I am referring to here are business customers or service providers who may have service solutions that are more complex due to multiple site delivery/network implementation or complex service requirements/solution or where the services require tailoring on a per customer basis. Therefore, the customer network and customer solution design should be part of the sales support process. These customer designs are normally carried out by technical sales personnel (e.g. customer network designer or technical sales consultants) and are agreed with the customers before customer pricing and contract signatures. An example sales engagement process is drawn out in Section 9.3.2. This example shows how the sales engagement fits within the downstream processes (i.e. the order management and service and network provisioning and billing processes) for the customer orders.

All the technical sales personnel will require technical training for the service. The following procedures/detailed work instructions are to be designed as part of the service design activities for services that involve designing the network solutions and tailoring of the service for each customer:

- customer leads tracking;
- customer qualification and credit checking;
- customer relationship management;
- customer information management, including what service they have bought and network CPE and site information;
- customer network order feasibility;
- customer network and service design.

9.2 Customer Service Processes

The customer/end-user service processes area includes all functions, tasks and events that customers/end users may initiate with regard to their service. The process requirements may vary between services; however, the operational processes in this area should cover (but are not limited to) the following topics:

- service enquiries;
- complaint handling;
- billing enquiries;
- customer technical support;
- service change requests (MACs);
- service and network termination and customer retention;
- customer/end-user migration;
- order management;
- fault management.

Customer service teams are the front line to the customers/end users. They project the image of the company and are the ones that form the customers'/end-users' perceptions of the service. Therefore, it is very important that they are well trained and equipped with all the right tools and information to serve the customers/end users.

When designing the customer service processes, one should bear in mind where/which communication medium (i.e. by telephone, e-mail or others) these enquiries or requests will be arriving from. To design a good service, one should look at it from the customers'/end-users' point of view and, for example, ask questions like:

- What is the preferred/most convenient medium for the customers/end users to access to the customer service/service request (e.g. the topical request areas are as list above)?
- Will the requests be arriving as e-mails?
- If it is by telephone, will it be a freephone number or are the calls to be charged at different rates?
- Will the service details and features listed on the website be enough to all answer enquiries?

Having the comprehensive service information on the website is always a good start, but from a customer/end-user point of view, psychologically, it is always better to talk to a real person for enquiries or requests. However, that could prove costly to the service. Keeping a balance between cost and service perception is needed as part of the service design activities.

Once the medium by which these enquiries/requests has been decided, one needs to ensure that they are channeled to the right operational teams for effective processing. If these enquires/requests are by telephone, then one needs to make sure the phone numbers are correct and route them through to the right customer services team according to the nature of the enquiries. As part of the service design activities, various scenarios and process flows within all of the areas listed above are thought through. Processes and actions to deal with the potential customer/end-user requests/enquiries should result.

The other important consideration is operational hours for the teams. Although longer operational hours will be more convenient for the customers/end users, this has huge implications on the operational cost.

9.2.1 Service Enquiries Process

The service enquiries process and work instructions should include all the potential questions that customers/end users may enquire about the service. Information regarding the service may include: what the service is; the unique selling points of the service; the service features and options and the combination of features and options where relevant; pricing on all the service features, options and combination of the service features/options; service provisioning lead times and any related SLAs.

Within this process, it may be a good idea to include information/guidance/instructions regarding upgrading/up-selling or cross-selling services to customers/end users. Customers/end users only normally enquire about services that they are interested in. Therefore, it is important that the customer service enquiry teams are trained to obtain and understand customers'/end-users' requirements to match the services they require. Training the teams to have good knowledge on all the related services is also important, especially when up-selling or cross-selling services.

9.2.2 Complaint Handling

The complaint handling process and procedures concern customer/end-user complaints. These complaints may result from a genuine mistake made by the service providers, but they can also be the result of a misunderstanding between the customers/end users and the service/actions on the service. Therefore, verification between the two is important. Dealing with complaints is one of the major areas that impacts on the customers'/end-users' perception of the service. If complaints are mishandled or customers/end users are not satisfied with the end result, then they are very likely going to change service providers.

As part of the service design activities, all the complaint scenarios should be considered (e.g. late service delivery, service request not performed and so on). For each scenario, work instructions are designed with actions to remedy the situation. If compensation is given to the customers/end users where mistakes are committed by the service provider, then the criteria for compensation should be detailed as part of the work instructions. The nature of customer/end-user complaints should be captured as part of the complaint processes. This information will be good feedback to improve the service and will be useful for future service enhancements.

9.2.3 Billing Enquiries

Billing enquiry processes deal with customers'/end-users' enquiries relating to bills, billing items and charges on their bills. Most queries are on the amount being charged or regarding calls that the customers'/end users' claim they did not make. Therefore, it is essential to have the customers'/end-users' service packages and pricing information, billing data or CDRs available to the bill enquiries team to ensure they have all the answers for the customers/end users. This ensures that all billing disputes can be resolved at the first instance. Procedures to deal with unusual/exceptional enquiries will also be required. These queries typically require further information and should be responded to within the time stated in the customers'/end-users' SLA.

Overcharging and issuing rebates for the service should be avoided where possible. Ensuring the billing accuracy of the service will eliminate many billing enquiries from customers/end users. However, detailed work instructions should include criteria for rebate where appropriate.

9.2.4 Technical Support

Technical support is of growing importance due the complex services and applications that are being provided. Providing a comprehensive technical support will definitely improve the customers'/end-users' perceptions of the service, but this can also be costly.

Not all technical enquiries are related to faults. Many are associated with initial service set-up (i.e. when the customers/end users first obtain the service and try to get it up and running) or educating the customers/end users on how to use the service. In an ideal world, all services are 'plug and play' services, where the end users just follow the instructions and everything will work first time. However, this is not often the case. As part of the service design activities, well-tested end-user instructions are important in order to avoid unnecessary enquiries. End-users' technical competencies are not guaranteed; hence, the instructions will need to be simple, well illustrated and comprehensive.

For business customer services, the volume of technical enquiries should be less, as the initial technical installation should be tested as part of the service provisioning process. However, the personnel supporting the business customers will require in-depth technical knowledge on the service, as the customers may have a fair amount of technical knowledge themselves, especially for customers who have an internal technical team that uses the service.

Information to be provided to the technical support team should include (but is not limited to):

- technical information on the service (e.g. how the service works end to end on a technical level, the technology used by the service);
- technical set-up for the service;
- customer/end-user equipment settings and configuration;
- potential symptoms of incorrect settings;
- list of technical questions that will help to diagnose customer/end-user problems;
- instructions to customers/end users to rectify equipment setting/configuration/problems.

Providing tools to enable the technical support team to have access to the customers'/end-users' equipment will be very useful to enable comprehensive technical support.

9.2.5 Order Management

Order management processes define the sequence of events required to handle all customer/end-user orders and order exceptions. It should also include the handling of customer/end-user order enquiries and order progression.

The order management process is initiated by the end users/customers or from the sales engagement process. The order management process ensures all the orders are progressing through the service and network provisioning process until order fulfillment is completed, with progress updates throughout the provisioning cycle for the orders. The process should also ensure the customers'/end-users' SLAs for service provisioning are met and that any breaches of SLAs or risk of jeopardy to the SLAs are reported.

As part of the order management process development:

- The customer/end-user order forms for the service are to be defined (including all the service features, available options and combination of service features/options).
- Items for order validation and validation criteria are listed (e.g. validation of end-user addresses).
- Checks are to be carried out before the orders are accepted (e.g. checks for duplicate orders).
- A mechanism is developed to ensure that customer/end-user accounts are created in customer/end-user management and billing systems.
- All the exception scenarios are to be listed with the required remedial actions to be taken.
- Order status and update points in the process are defined. Examples of order status may include: open – order being validated/processed; awaiting network resources; progress to provisioning; closed/completed;
- A mechanism and process for informing the customers/end users of their order progression are defined.
- The handover points and information passed between the order management process and the service and network provisioning process are identified.
- Escalation procedures are put in place for orders that are about to breach and those that have breached the customers'/end-users' service provisioning SLAs.

It is important to perform as much validation of the orders as early as possible, preferably before accepting them. This will reduce the risk of order failures and the number of order exceptions further down in the order processing life cycle.

For end-user orders, most of the process/tasks listed above, except for the escalation procedures, can be automated (or should be automated as much as possible). The order management process for customer orders may involve more manual procedures for all or some of the tasks above. This is due to the potential complexity of the customer service solution. Customers that are being referred to here are business customers or service providers who may have services solutions and network implementations that are more complex due to multiple site delivery or complex service requirements/solutions. The process for providing a complex service solution to customers normally starts with sales engagement where the customers' requirements are captured and the customer solution agreed. The order management process for customers starts after contract signature with the customer. After validation, the orders are passed over to the service and network provisioning process. The order management process ends after the orders have been completed with service tested and billing triggered. The progression of the orders is monitored through the order management process.

The end-to-end order management and service and network provisioning processes are illustrated in Section 9.3. It is useful to develop a high-level end-to-end order management and service and network provisioning process, as this shows how the orders are completed from start to finish. Having the end-to-end process gives everyone a view of the process flow, especially as an end-to-end process spans multiple operational functions/areas. The process should also show the handover points and the responsibility demarcations involved for each operational department/functional area.

9.2.6 Service Change Requests: MACs

The service change request process starts with the definition of the different types of service change request that are available as part of the service. Change requests for a service may include (but are not limited to):

- changing the billing address (excluding moving house/moving customer sites);
- changing customer/end-user service options and packages;
- changing an end-user's name on the bill (due to change in personal circumstances or it was wrongly captured during the end-user creation process);
- adding new/additional network equipment at the customer site;
- adding a second network connection to the customer site;
- adding a new customer network connection to the service;
- moving house for the end users;
- moving customer network equipment to a different location within the same building;
- moving site (i.e. moving network equipment to a different geographic location).

For each of the service change request scenarios, information required (to perform the request) from the customers/end users is captured (usually in form of a request form) and operational support processes to handle these scenarios will need to be defined. These processes may be linked to other operational support processes, like service provisioning processes. For example, the change of service options/packages can be linked into the service provisioning process. Adding a second network connection to the customer site may be treated as a new network connection provisioning process.

Operational processes for business customers can be very different from those of end users, as the network connections are more expensive to move/change and will may require a longer lead time. This is especially true for fixed-line services. For each service change request requiring moving or adding a new site for a business customer, a short feasibility study is advised to ensure that the change can be carried out successfully without affecting the rest of the service. As part of the feasibility study, the pricing/cost of the change should result. Pricing for change requests should also be provided to the customers/end users before carrying out the changes to their service.

9.2.7 Service and Network Termination, Cancelation and Customer Retention

Service and network termination and cancelation processes deal with customers'/end-users' requests to terminate/cancel the service or business customers' requests to terminate network connections. All service and network termination requests are initiated by the customers/end users. As mentioned in Chapter 8, network termination does not necessarily equate to the termination of the service, especially in the business customer environment, where the customer may close one of their sites. This does not equate to terminating the whole service. This is equally true for service termination. Terminating a service does not necessarily mean the termination of the network connections, as multiple services may run over the same network connection, although the bandwidth of the network connection may need to be adjusted in light of the service termination.

The service and network termination/cancelation process can be split into two parts:

- receiving and verifying the requests from the customers/end users;
- performing the service and network termination requests.

For most organizations, the customer service processes will cover the customer interface activities, whilst the actual termination requests will be carried out by the service provisioning area. Under this scenario, it is useful to have a high-level end-to-end process to ensure that the activities are complete and that the handover points and information exchanged between different operational areas is clearly defined. Detailed processes and work instructions for each stage of the process will result from the agreed high-level end-to-end process. An end-to-end service termination process example can be found in Section 9.3.

As part of the development of the service and network termination process:

- customer/end-user request forms (with all the necessary information required to terminate a service/network connection) are defined;
- request validation/items for validation and validation criteria are listed (e.g. whether the customer/end user currently on the service is authorized to terminate the service);
- all the exception scenarios and request rejection scenarios are listed with remedial actions required to be taken;
- the progression status and update points in the process are defined;
- mechanism and process for informing the customers/end users with their request progression are defined (if required);
- the handover points and information passed between the two parts of the internal processes for service and network termination are identified; please see Section 9.3.3 for further details.

9.2.7.1 Cancelation

As part of the cancelation process, it is essential to confirm that the customers/end users are eligible for cancelation. If cancelation/termination incurs additional charges, then end users/customers must be informed and agree with the charges before the requests are to be accepted. Cancelation charges can occur if the end users are to cancel the service after the eligible cancelation period or if the end users want to terminate the service within the minimum term of the service.

9.2.7.2 Customer/End-User Retention

As part of the termination process, it is a good idea to put in place procedures to capture the reasons for cancelation and termination. Scenarios of customers'/end-users' terminations should be listed with various pricing incentives and up-selling strategies to retain the customers/end users. Customer service teams are to be trained with strict guidelines on what can be offered to the end users under what scenario. Authorization on what can be offered should be clearly stated. This will avoid unnecessary and costly mistakes. A procedure to analyze customer churn may also be included in this area.

9.2.8 Customer/End-User Migration

The customer/end-user migration process deals with customers/end users that are being migrated from one service to another. This can be customer/end-user driven as well as service provider driven. For example, the customer may want to migrate from the current broadband service to a leased-line service for several of their sites due to rapid expansion plans and resiliency requirements. Or the end users may want to move from their second-generation mobile service to the 3G service. Or the service providers decided to withdraw a loss-making service and require migrating their customers/end users to a more profitable, value-added service.

For fixed-line business customers, service migration may involve service feasibility and coordination of resources to ensure all the equipment, field engineers and the network configurations for the customers are changed at the right time and in the correct sequence.

Migration processes are mainly used when there the current service is being withdrawn. Customer/end-user migration processes are normally defined as and when there are migration requirements, rather than as part of the service launch, unless the new service is superseding a current service or it has been identified that there is a high chance that customers/end users may change to the new high value-added service after service launch. The customer/end-user migration processes should be defined in conjunction with the service withdrawal or service migration processes. Please see Chapter 12 for details relating to service withdrawal and service migration.

9.2.9 Customer Fault Management

The fault management process referred to in this book is about handling faults reported by customers/end users. It is also known as the trouble ticketing and the problem resolution process. The main differences between the faults reported by business customers and end users are that the business customers normally have a tighter SLA for fault resolution and that the volume of faults reported is less. From the process point of view, handling both is much the same, but the scale of operation and system requirements to support the fault management operation may be different due to the potential volume of faults for end users and the complex network set-up for business customers. For end-user faults, the service is normally provided over one network connection; therefore, fault diagnosis will be simpler. Hence, automation for fault diagnostic tools should be developed where possible. This type of automation is essential to achieve efficiency and cope with the volume of potential faults. The networks for business customers are more complex; hence, more sophisticated diagnostic tools with more detailed work instructions and training/skills are required to resolve these faults.

When defining the fault management process, one needs to bear in mind where the fault tickets will be coming from and what mechanism the customers/end users can use to raise these fault tickets. The estimated fault volume is also very important to the size of the operational the team and, hence, to the operational cost of the service. The opening hours and the fault resolution SLA for the service also have a huge impact on team size, operational hours and shift work requirements.

Within the fault management process, there should be (but this is not limited to):

- Diagnostic questions for the service to capture fault symptom details and to perform initial diagnostics. Diagnostic questions should be defined as part of service design activities. The more detailed the diagnostic questions, the easier it is for the support team to pinpoint the problem. Diagnostic questions should be in the form of a flowchart with different questions depending on the symptoms of the problem. The answers to the series of diagnostic questions should help with pinpointing the problem area.
- Detailed instructions for the use of any diagnostic tools (if any). The instructions should include: how the tool can be used; which tests within the tool should be applied for which symptoms; and the interpretation of the test results.
- Process feeds from the network management process for planned engineering works and current open faults (with estimated fault resolution time) to avoid opening unnecessary fault tickets or duplicate tickets for the same fault condition. This will also assure customers/end users that the problem is known about and being dealt with. An estimated fix time should be given to the customers/end users if known.
- Process feeds to the network management process for fault diagnosis and fault resolution (if required).

- Roles and responsibilities for each of the fault-handling operational teams; detailing which teams are responsible for handling the faults for the service. Clear demarcation of responsibilities should be stated and handover points with a defined set of the handover information required.
- Procedure to raise the fault tickets, including the minimum information required when raising a fault ticket (e.g. customer name, contact person, fault location, fault symptoms and diagnostic tool test result).
- Procedure to provide fault resolution progress updates to customers/end users. This depends on the service requirements for the service.
- Detailed work instructions to perform root cause analysis on faults on the service and network equipment used for the service.
- Procedure to log network fault with OLO providers if it is found that the fault is with their network (if relevant).
- Definition of fault severity categories. These severity levels are normally tied with customer SLAs relating to fault resolution times. Example fault severity categories may include:

  o *Category 1* – total loss of service (fault resolution SLA of 5 h);
  o *Category 2* – partial loss of service (fault resolution SLA of 10 h);
  o *Category 3* – degradation of service (service to be restored within 24 h);
  o *Category 4* – minor fault, not affecting service (corrective action to be applied within three working days).

- Defined fault status and points of fault status updates within the fault management process. Example fault status include:

  o open; no fault found; await OLO resolution; further investigation required; await engineer to be on site to resolve fault; closed.

- Procedure to handle fault tickets that are about to breach the customer fault resolution SLA.
- Escalation procedures for fault tickets that have exceeded the fault resolution SLA.
- Procedure to monitor the performance of external fault resolution agencies (if used). External fault resolution agencies can be OLO network fault resolution team, third-party supplier who provides the equipment required to fix the fault or a subcontractor used to go to customer sites to fix the faults.

It is useful to have a high-level end-to-end fault management process, as this process often spans different operational areas. Having the end-to-end process will give everyone a view of the work flow and will show the handover points for each operational area. The demarcation of responsibilities within the fault management process depends very much on the operational organization of the company. There is no strict rule about this. In the example below, I have divided the responsibilities based on the technical skills required to perform the tasks. The example illustrates the sequence of events and tasks to be performed as part of the fault management process, and the demarcation of responsibilities is for reference only.

9.2.9.1 Fault Management Process Example

Figure 9.1 is an example of a fault management process. The example is for a managed IP-VPN service for business customers. This should be of a high enough level to be nontechnology specific, but detailed enough to have the concept applied to most managed services. The description of the process is given in Table 9.1.

The fault resolution team can be an internal second-line support team who are authorized to perform actions to remedy the fault or field engineers at a switch site/customer sites to resolve the

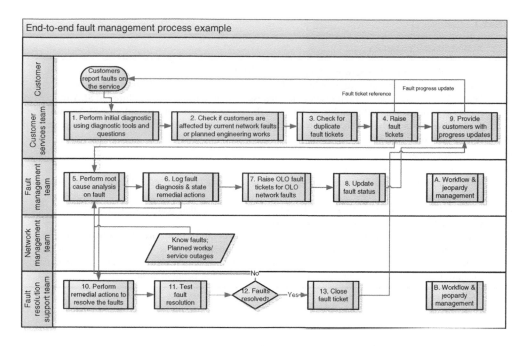

Figure 9.1 Fault management process example.

Table 9.1 Description of a fault management process.

Steps Actions/tasks description

Customer services team

1   Obtain answers for initial diagnostic questions to narrow down the area that might be faulty. Where relevant, the diagnostic tools (e.g. performing 'ping' test to establish basic network connectivity) should be used.

2   Check if the customer is affected by existing faults or is part of any planned engineering works.
    Inform the customer of estimated time of fault resolution if the customer is impacted by current engineering works or faults that are already known and are being resolved.

3   Check if customers already had a fault ticket open with the same fault.

4   If it proves to be a genuine problem after checking for known problems, going through the diagnostic questions, and diagnostic tool test results, then fault ticket will be raised. As part of raising the fault ticket, a fault severity category is assigned. Customer is given the fault ticket reference.

9   Provide customers with an update. Depending on the service requirements, some services may require the customers to be updated with the progress every 2 h.

Fault management team

5   Perform fault diagnostics. This may include: checking if network alarms have been raised against the network equipment used to provide the customer with the service; checking if planned works exist for the network equipment involved; check if another fault ticket has already been opened for the same fault, but is still being resolved; performing routine checks on the network equipment involved; logging onto the customer site router to check configuration; and performing any root cause analysis where necessary, and so on.

6   The fault diagnostic should be logged and actions to remedy the situation should be stated. Diagnostic tests and test result should also be logged as part of the fault resolution on the fault ticket. The fault diagnostic result could be: no fault found; fault diagnosed (stating the root cause), await engineer to be on site to fix the fault; further diagnostics required.

(*continued overleaf*)

Table 9.1   (*continued*).

Steps	Actions/tasks description
7	If the fault lies with OLO network, fault tickets with OLO are raised.
8	Fault ticket should be updated to reflect fault ticket status in light of the diagnostic results.
A	Work flow management of fault resolution progression should be monitored. Internal SLA trigger points at various steps of the end-to-end process should be set-up to ensure customer SLA is not jeopardized. If the fault ticket is about to go into jeopardy, alarm should be raised to the management for more attention.

Fault resolution support team

10	Carry out fault resolution actions.
11	Test the network and service after fault resolution actions have been carried out to ensure the faults have been resolved.
12	Check if the faults have been resolved. If they have not, further diagnostics will be required.
13	Close fault tickets when faults have been resolved.
B	Work flow management of fault resolution progression should be monitored to ensure customer SLAs are not jeopardized. SLAs with third-party fault resolution agencies (e.g. OLO fault resolution team or subcontractors for field engineering work) are measured. If the fault ticket is about to go into jeopardy, escalation process should be invoked to prevent customer SLA being breached.

fault or third-party suppliers; if it is found to be an OLO network fault, then it is the OLO's responsibility to resolve the fault within their SLA.

9.3 Service and Network Provisioning

Service and network provisioning processes activate the customers and the end users on the service. As mentioned in Chapter 8, service provisioning is not the same as network provisioning, especially in this telecommunication era where multiple services can be delivered over one network/network connection. I have also separated the service and network provisioning of end users from that of customers (by customers, I mean business customers or service providers who have a complex service solution and/or a multiple site network delivery). There are distinct differences between provisioning the customers and the end users. This is mainly due to the complexity of the service solution involved – the complexity lies within the customer network and service solution design, the network implementation of multiple site locations (i.e. multiple access network connections) with different configurations for services on the CPE for each site and the coordination and project management skills required during the service and provisioning process. Hence, more manual labor and operational procedures are required. Where possible, the processes should be automated. For end-user provisioning, services are implemented over one access network connection (whether it may be mobile or fixed network connections); hence, it is much simpler, but the challenge will be on the volume of orders to be provisioned. Therefore, the end-user service and network provisioning process should be automated as much as possible to achieve operational efficiency.

9.3.1 End-User Service and Network Provisioning

The end-user provisioning process is the next stage and a subset of the order management process described in Section 9.2.5, where end-user orders have been accepted after validation. The end-user service and network provisioning tasks will vary depending on the service involved; however, the basic conceptual process can be as below and should contain the following procedures/detailed work instructions or system process (if automated) to ensure the actions are carried out correctly:

1. Define end-user service and network order after investing in the network resources required. The definition should contain all the network elements to be provisioned and the appropriate service profile to be applied for each end-user order.
2. Check network resource capacity. If there is a lack of capacity or the capacity is low, then the capacity management team should be informed. This action should be linked to the capacity management process. If there is no capacity, then the end-user orders should be put on hold until the necessary network capacity becomes available. This is one of the exception scenarios. The definition of actions required for this should be defined as part of the service design activity, as this varies between services. Things to consider will primarily be the end-user SLA and whether the end users are to be informed.
3. Order end-users' CPE (if required). This can be done as part of the order management process, depending on whether the CPE requires network resources allocation.
4. If network resources from OLO are required, then orders for these network resources are created and sent to the OLO. The status of OLO orders should be monitored. OLO order failures will become provisioning exceptions, as the end-user orders cannot be completed.
5. When all the network resources are available, end-user network provisioning activities can take place. The actions for this are dependent on the network technology involved and should be automated to avoid human error and to improve efficiency.
6. After the network resources have been successfully provisioned for the end users, the end-user service profile is applied to the relevant network resources. The definition of the service profiles/policies and the network parameters to be applied are dependent on the service. The service profiles should be defined as part of the service design activities. Please see Section 8.8 for further details. The service profile provisioning task should also be automated to avoid human error and to improve efficiency.
7. The network inventory and service profile database should also be updated with the end-user provision details.
8. If it is a managed service, where end-user CPE is monitored, then the network management system will need to be provisioned with the end-user network and service details.
9. Similarly, for a managed service, the end users may have management reports as part of the service, so the end users and their network and service details are to be provisioned in the relevant reporting systems.
10. Once the network and service have been provisioned, end-to-end testing of the service should be performed where possible. This is to ensure that the service is working and to avoid end-user complaints.
11. Upon successful service testing, billing of the end users is triggered with the service provisioned date. The appropriate charges are applied and the end users are billed as part of the billing process. If the service is a prepaid service, then the service profile should define the amount of credit the end users have for the service and the billing trigger will probably start when the end users make the first call/log-on session. Credit will be deducted from then on.

An example end-user service and network provisioning process is illustrated below.

9.3.1.1 End-to-End Process Example for End-User Order Management and Service and Network Provisioning

Figure 9.2 is an example end-user order management and service and network provisioning process for an end-user broadband service. For each of the steps shown in Table 9.2, detailed work instructions/system processes should be written to ensure they are carried out in the correct way.

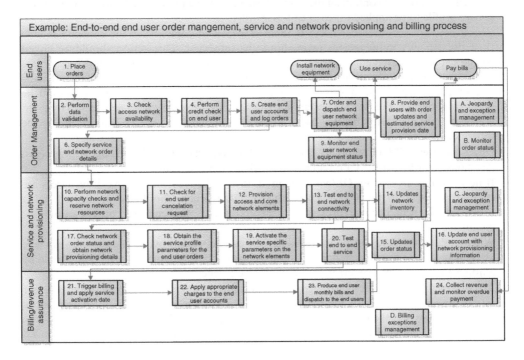

Figure 9.2   End-to-end process example for end-user order management, service and network provisioning and billing process.

Table 9.2   Description of an end-user order management and service and network provisioning process.

Steps	Actions/tasks description

Order management team

1	End users decided to place broadband orders.
2	Validate data in the end-user orders. Data to be validated includes: end-users' names, addresses and phone numbers and valid service package/profile.
3	Check broadband access network availability and capacity for service at the end-user's address. For a broadband service, the access network availability checks are carried out at an early stage of the order life-cycle because that is the part that causes most order failures.
4	Perform credit check on end user to ensure they do not have any bad debts.
5	After successful checks and validations (steps 2–4 above), end-user accounts can be created (in end-user management system and billing system) and the orders are created and logged (in the order management system), ready for processing. Creating end-user accounts in the billing system may include the setting up of direct debts for the end users for payment collection. Direct debit authorization and bank account details from end users are to be obtained.

Table 9.2 (*continued*).

Steps	Actions/tasks description
6	Using the end-users' order details, specify the relevant network elements (access and core network) that require provisioning and the service profile for the end-users' services.
7	This broadband service includes the end-users' network-terminating equipment (CPE). Hence, the equipment is to be ordered and dispatched to the end users.
8	Provide end users with order progress updates and estimated time of order completion.
9	The dispatching of the end-user equipment is monitored to ensure it arrives when the service is meant to be available. Please note that steps 6–9 are performed in parallel to improve efficiency.
A	At any step above, the order may fail. Jeopardy management processes are in place to deal with all failure scenarios. It should be possible that the orders may be fixed manually and put back in the order management system for further processing.
B	Work flow management of order progression should be monitored. Internal SLA trigger points at various steps of the end-to-end process should be set up to ensure end-user provision SLA is not jeopardized.

Service and network provisioning team

10	Check if all the network resources (e.g. access network ports) are available for each end-user order. Reserve and allocate network capacity where appropriate.
11	Check if the end users have canceled the orders before provisioning the network elements.
12	Provision all the network elements (e.g. access network ports, core network resources)
13	Perform end-to-end network connectivity testing where appropriate. This is only possible if the end-users' network equipment has arrived at the end-users' premises and is plugged into the network termination point and working. Advice should be given to end users with regard to this.
14	Update network inventory with the new CPE at end-user site and network provisioning information. This also includes the update on network resources used relating to capacity availability of the network.
15	Update order status in the order management system.
16	Update end-user accounts (in the end-user management system) with network provisioning information. This is used in the service provisioning process and it will also be useful for fault diagnostic purposes.
17	Check the network elements have been provisioned and that the end-to-end network connectivity tests have been successful.
18	Obtain the required network parameters (e.g. QoS parameters, site/call/session authorization mechanism) from service profile database.
19	Activate the service-specific network parameters on the appropriate network elements and update the order status.
20	Where possible, perform service testing for end-users' services. Please note that tasks 14, 15, 16 and 17 can be performed in parallel.
C	At any step above, the network provisioning order may fail. Jeopardy management processes are put in place to deal with all failure scenarios. It should be possible that the orders are fixed manually and are put back in the order management system for further processing.

Billing and revenue assurance team

21	Trigger billing for the end users with the service activation date upon successful service testing.
22	Apply appropriate charges to the end-user accounts.
23	Produce and dispatch monthly end-user bills in the appropriate billing cycle.
24	Collect revenue that is due and monitor overdue payments.
D	Billing exceptions from the billing processes will need to be attended to by billing operational teams. List of possible billing exceptions and their remedial actions are to be defined as part of the service design activities.

9.3.2 Customer Service and Network Provisioning

The customer service provisioning process starts after the customer orders have been validated. The provisioning process will vary between services, as service-specific requirements are to be catered for. However, the basic conceptual steps can be as follows:

1. Review customer network and service solution design.
2. Identify network resource requirements for the customer service solution.
3. Check network capacity availability for the required network resource. A network feasibility study should have been performed as part of the customer network design. Therefore, any capacity shortfall should have been identified. Where there is little or no capacity, additional capacity (e.g. additional access ports) should be ordered. These requirements should be fed into the capacity management process and should be dealt with in conjunction with the capacity planning and network build team, as the additional capacity required may already have been planned or is being delivered.
4. A detailed network connectivity design is produced showing how the customer network is to be connected through the network. This includes details like the network ports to be used and logical connections through the network.
5. Order CPE.
6. If part of the network (e.g. access network) is to be provided by other network providers (OLOs), these network connections are ordered.
7. When the delivery dates of steps 3, 5 and 6 are known, coordination with the customers on site installation dates are required, especially if access to customers' sites is required and if it involves multiple-site delivery. Resources for on-site installation are arranged.
8. After successful access network provisioning, core network provisioning takes place.
9. On-site installation and configuration of CPE. Network connectivity testing is carried out as part of the site installation.
10. Update network inventory with the customers' CPE details and network provisioning details.
11. Upon successful network provisioning and installation, service provisioning takes place. Service parameters or a service-specific configuration is applied to the relevant network elements.
12. Depending on the service, customer acceptance testing may be performed. This is the formal sign-off by the customer that the service has been implemented successfully.
13. Update order status.
14. Billing is triggered after customer acceptance testing is successful and when the customers have accepted the service.

For a complex delivery, a project plan for network and service implementation for all customer sites will be required. Detailed work instructions for steps 2–11 for the service will need to be designed as part of the service design activities. In the section that follows is an example end-to-end design and provisioning process for an IP-VPN service. It is intended to provide a guide as to the steps and sequence of events that takes place from the sales engagement activities through to service and network implementation.

9.3.2.1 End-to-End Process Example for Sales Engagement, Customer Order Management, Service and Network Provisioning and Billing Process

The example in Figure 9.3 is the end-to-end sales engagement, order management, service and network provisioning process for an IP-VPN service. This service is chosen because of its complexity to illustrate the various potential dependencies. For customer orders, the end-to-end process starts at sales engagement; see Table 9.3.

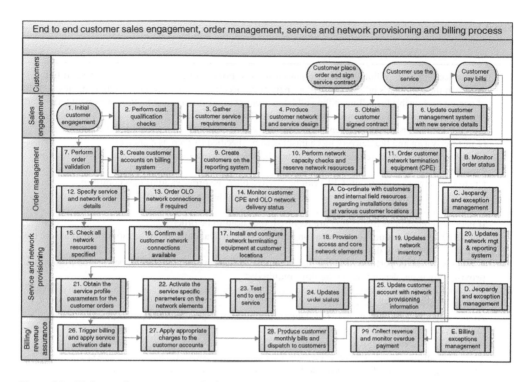

Figure 9.3 End-to-end process example for customer sales engagement, order management, service and network provisioning.

Table 9.3 Description of sales engagement, customer order management, service and network provisioning and billing process.

Steps	Actions/tasks description

Sales engagement team

1   Customer engagement can be initiated through sales leads or from existing customer relationships, as well as by responding to tenders.

2   Upon satisfactory customer qualification checks, create customer on customer management system.

3   Develop customer relationships and gather customer requirements.

4   Produce customer network and service solution design document in accordance with their requirements. As part of the customer network and service solution design, a network capacity feasibility study is performed. The feasibility study should detail the network cost for the customer's network.

5   As part of the customer contract signature, pricing for the service is agreed, an order form for the service is completed and the customer network and service solution design is signed off. These are logged in the customer management system.

6   Update customer account in the customer management system with this service order.

Order management team

7   Order validation includes: verification of customer contract details, ensuring that the contract has been signed with the right level of authorization; verifying the credit history of the customer; checking if this is an existing customer; checking that all the customer information (e.g. billing address, site addresses, name of person responsible for various touch points like fault escalation and billing) is present; checking that the customer network and service solution design are complete and signed off by the customer; checking that all the mandatory fields on the order form have been filled in.

(continued overleaf)

Table 9.3 *(continued).*

Steps	Actions/tasks description
8	Create account in the billing system for new customers. For existing customers, create the additional service in the customer's billing account. As part of the account creation, company name, name and title of person responsible for billing, billing address, account number, payment method, payment terms and pricing information of this service are entered into the system.
9	In order for the customers to receive service management reports, accounts are to be created in the reporting system and the delivery mechanism is to be specified (if there is a choice).
10	Identify all access and core network ports and check for network capacity. Reserve network resource/capacity if required.
11	Customer network equipment (CPE) can be ordered once the order validation is complete.
12	Network and service order details are specified, including the access network locations and service profile requirements.
13	If network connections from OLO are required, they need to be ordered.
14	The OLO network connection installation date and CPE arrival dates are to be monitored to ensure the customers' installation dates are kept.
	Steps 8–12 can be performed in parallel to improve efficiency. Step 12 may be performed by network provisioning team, depending on the skills required.
A	Coordinate with customer on installation dates. This can only happen when all the installation dates and equipment arrival dates are known.
B	At any step above, the orders may fail. Jeopardy management processes are put in place to deal with all failure scenarios. It should be possible that the orders may be fixed manually and put back in the order management system for further processing.
C	Work flow management of order progression should be monitored. Internal SLA trigger points at various steps of the end-to-end process should be set up to ensure customer provision SLA is not jeopardized.

Service and network provisioning team

15	Check all network resource (e.g. all access and network ports and routing around the core network) that have been identified (on a per network connection basis) are correct (i.e. per customer network design) and the network capacity reserved where necessary.
16	Confirm all access network connections and network resources are available.
17	Install and configure network equipment at customer site in accordance with the network design.
18	Provisioning all the network elements (e.g. access network ports, core network resources). This can be done only when steps 15–17 above have been completed.
19	Network inventory is updated with CPE details and network provisioning information.
20	(a) This is a managed service. As part of the provisioning process, the CPE is provisioned on the network management system for monitoring and management purposes. (b) Reporting systems are to be activated in order for the system to collect management data and compile reports for the customers.
21	Obtain the required network parameters (e.g. QoS parameters, site/call/session authorization mechanism) for the service from service profile database.
22	Activate the service-specific network parameters on the appropriate network elements and update the order status.
23	Perform end-to-end network and service testing for all the customer network connections, ensuring the network connections to all the sites are working as designed and that the service required is available for all customer locations.
24	Order status is updated.
25	Customer account in the customer management system is updated with the network provisioning details. This will help with customer enquiries.
D	Workflow management of network and service provisioning should be monitored. Internal SLA trigger points at various steps of the end-to-end process should be set up to ensure customer provision SLA is not jeopardized.

Table 9.3 (*continued*).

Steps	Actions/tasks description

Billing and revenue assurance team

26	Trigger billing for the customer accounts with the service activation date upon successful service testing.
27	Apply appropriate charges to the customer accounts.
28	Produce and dispatch monthly customer bills in the appropriate billing cycle.
29	Collect revenue that is due and monitor overdue payments.
E	Billing exceptions from the billing processes will need to be attended by billing operational teams. List of possible billing exceptions and their remedial actions are to be defined as part of the service.

9.3.3 Service and Network Termination/Cancelation

The process steps for dealing with service and network termination/cancelation requests are the reverse of the provisioning process. Essentially, you are deprovisioning what has been provisioned in the service and network provisioning process. Most service and network termination/cancelation requests come from the customers/end users. However, some service termination requests may come from the billing/finance department when the customers/end users have not paid their bills. The termination process is the same for both, but processes to activate such a termination will be required (please see Section 9.8 for details). Items/tasks/ procedures to be included as part of the service and network termination/cancelation processes are as follows:

- Check that the service and network to be terminated/canceled have been provisioned.
- If no network/service resources have been provisioned, then cancel the order in the order management system. Log the cancelation request and no further action is required.
- Define all the network elements (including OLO network resources) to be deprovisioned.
- Obtain the appropriate service profile of the service to terminated/canceled without network termination.
- Apply the appropriate service profile/delete the appropriate service profile from the appropriate network resources.
- Deprovision the network resources.
- Release network and system resources for reuse.
- Update status at appropriate process points.
- Provide termination updates to billing system for final bill production for the end users/ customers.

All the above tasks/procedures can be applied to both end users' and customers' service and network termination. An example process for customer service and network termination is provided in the following section.

9.3.3.1 Customer Service and Network Termination Example Process

For the business customer, the service and network termination process will be more complicated than that for end users. This is likely to be a project-managed activity. Figure 9.4 illustrates an example process for an IP-VPN service. Customers initiate service termination either through the customer account team or for small customers, through the customer service team; see Table 9.4.

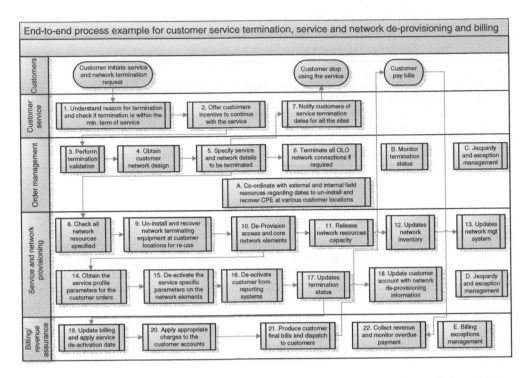

Figure 9.4 End-to-end process example for customer service termination, service and network deprovisioning and billing.

Table 9.4 Description of customer service termination, service and network deprovisioning and billing process.

Steps	Actions/tasks description
Customer account or customer services team	
1	When customers initiate service termination requests, it is essential to understand why they are leaving the service. At this stage, the process should prompt to check if the customers are terminating the service within their minimum term of the contract. If that is the case, it should be highlighted to the customer and charges are involved with the termination.
2	Incentives to retain customers should be used.
7	Once the dates for network termination of all network elements are known and field resources are booked to uninstall and recover the CPE, the customers are consulted on the uninstallation dates and decommissioning arrangements are agreed with the customers.
Order management team	
3	Validation of customer service termination request should include: if the customers exist and that the services they are terminating are valid; validate if minimum term of the service contract has been fulfilled; check all the mandatory fields in the termination request forms have been filled in.
4	Obtain the customer design documents from the customer management system.
5	From the network design and the network inventory, specify all the network elements that require deprovisioning. From the service profile database, all the service profiles to be deactivated should also be specified.
6	Send network termination request to all OLO network connections (if any).

Table 9.4 (*continued*).

Steps	Actions/tasks description
A	Coordinate with external and internal resources and agree decommissioning dates for all the sites with customers.
B	Work flow management of service termination progression should be monitored to ensure that all the tasks are progressing and that they will be completed by the dates committed to the customers.
C	At any steps above, the order management tasks may fail. Jeopardy management processes are put in place to deal with all failure scenarios. It should be possible that the orders may be fixed manually and put back in the order management process.

Service and network provisioning team

8	All the network resources to be deprovisioned are checked to ensure that they are accurate and that nothing is missing.
9	All the CPE at customer sites is uninstalled/decommissioned and the equipment recovered for future reuse.
10	All access and core network resources for that customer are deprovisioned.
11	Network capacity for the deprovisioned network resources is released.
12	Network inventory should be updated after the network resources have been deprovisioned.
13	Network management systems are updated, removing the deprovisioned network elements from management.
14	Obtain the required network parameters (e.g. QoS parameters, site/call/session authorization mechanism) from service profile database.
15	Deactivate the service specific network parameters on the network elements.
16	Customer reporting systems are updated, removing the customers from the reporting system and disabling any access to the system. All customer reports should be archived.
17	Status of network and service deprovisioning/termination is updated.
18	Update customer account in the customer management system with the service termination date.
D	At any step above, the service and network deprovisioning tasks may fail. Jeopardy management processes are put in place to deal with all failure scenarios. It should be possible that the termination requests can be fixed manually and put back in the service and network deprovisioning process.

Billing and revenue assurance team

19	Update customer billing accounts with the service deactivation date upon successful service termination.
20	Apply appropriate charges to the customer accounts.
21	Produce and dispatch final bills for the customers.
22	Collect revenue that is due and monitor overdue payments.
E	Billing exceptions from the billing processes will need to be attended to by billing operational teams. A list of possible billing exceptions and their remedial actions is to be defined as part of the service.

9.4 Service Management Processes

There are two types of service management processes: proactive and reactive. Proactive service management processes aim to manage the service such that the customers/end users will not notice any degradation in service performance and that the customer/end-user complaints are kept to the minimum. As part of the service design activities, the following proactive service management processes should be defined:

- Processes for monitoring and the management of internal service KPIs.
- Processes for monitoring customers;/end-users' SLAs.
- Process to conduct service operational reviews in light of service KPI results and customer/end-user SLA breaches to address possible improvements and issues.

- Process to define and implement service improvements as a result of customers'/end-users' feedback.
- Process to provide customer/end-user management reports.
- Process to deal with customer/end-user management reporting exceptions.
- Process to provision service KPI on the reporting system/systems for the respective functional/operational areas.
- Process to define and introduce new KPI for the service.
- Detailed work instructions to use the reporting systems available or the reporting functions of the systems in the respective functional/operational areas.
- Detailed work instructions to generate ad hoc reports from the reporting tools/system or to use reporting tools.
- Informing customers of planned engineering work or faults that affect service.

Reactive service management processes define how the customers/end users and the service should be managed when unforeseen events occur (e.g. occurrence of service-affecting network or system faults or dealing with customer requests that are not part of the defined service). The following processes/procedures/detailed work instructions should be considered as part of the reactive service management processes:

- Process to handle breaches of customers'/end-users' SLA.
- Detailed work instructions to assess customer/end-user impact regarding network and system faults using the available service management and network management tools. This is sometimes done as part of the network management process, depending on the operational organization and the service information availability.
- Process for informing customers of planned engineering work and service-affecting network faults.
- Detailed work instructions for the authorization and the application of pricing adjustments, discount, rebates and issuing of service credits on the billing system due to service complaints, breach of customer SLAs or operational errors.
- Detailed instructions on explaining the meaning of the SLA to the customers, why they were breached (if at all) and actions taken for future improvements.
- Escalation process to handle unresolved service issues.
- Process for dealing with customer service requests for complex solutions.

For services involving complex service solutions or for customers who have bought multiple services (hence, they are important customers from the revenue generation point of view), a service account management function is set up to deal with those accounts exclusively to ensure customer satisfaction and to make them feel that they have been served. The responsibilities of these service account personnel are to ensure that:

- the customers are kept informed with faults affecting their service and that these faults are resolved within the contracted SLA;
- the customer service requests are dealt with and the customers are happy with the progress of these requests;
- any service-related issues are addressed and resolved in a satisfactory manner.

They also act as the customer's advocate to ensure the customer's voice is heard and obtain resources to perform a customer's service-related tasks or requests. The aim of their actions should be to ensure that the customers are delighted with the service provided.

9.5 Network Management and Maintenance Processes

Network management and maintenance processes are designed to ensure that the network is operating at its optimum and that any network events are attended to. The aim is to manage the network proactively and detect and fix the network problems before service degradation is noticed by customers/end users. Areas to be covered should include:

- network monitoring (including the management of fault resolution agency and resources);
- network performance management;
- network disaster recovery;
- network configuration management;
- network maintenance.

9.5.1 Network Monitoring

Monitoring the health of the network is a major part of network management. Actions are to be taken when network events/alarms have been raised from the network elements to the network management system. The remedial actions for each alarm type should be defined as part of the network design activities. Within the network monitoring process, detailed work instructions for the following should be included:

- The checks to be carried out for initial diagnostics when network alarms appear on the network management system.
- Opening and closing trouble tickets on the trouble ticketing/fault management system. Trouble tickets are opened when a fault is detected. This is to ensure that the fault resolution is tracked. The minimum set of information to be included as part of the fault/trouble ticket should be defined (e.g. network equipment ID, description of the problem, severity levels, services and customers affected).
- Selecting trouble ticket severity categories. The category definition can be the same as that used by the fault management process for customer-/end-user-reported faults. Severity category can also be based on the number of customers/end users affected by the network fault. The fault-resolution SLA for each category should also be stated. The fault-resolution SLA is usually defined in conjunction with the SLAs.
- Prioritizing the fault diagnostics and resolution. There will be times when many network events and alarms arise at the same time. They need to be dealt with in accordance with the severity of the alarms and the service degradation associated with the network fault.
- Problem diagnostics and troubleshooting for the network elements used in the service. This should include a list of diagnostic tests to be applied for the different symptoms or alarm types. Corrective actions and fault resolution agencies for each potential network problem should also be recorded on the trouble tickets
- Remote log-on to network equipment with security precautions.
- Passing fault details to the fault management and service management teams/processes. This should include details like network fault severity levels, service and the customers impacted, assessment of the extent of service degradation and estimated fault resolution times.
- Trouble ticket status definition. Example trouble ticket status can be: open – await investigation; problem diagnosed – await to be fix (stating timescale); close.

- Testing to be performed for each of the fault conditions after fault resolution actions have been applied.
- Management of fault resolution agencies and resources. This could be internal resources (i.e. field engineers) or third-party support agencies (e.g. suppliers) or OLO network support. The purpose is to ensure that they perform the tasks/requests within the agreed SLA.
- Monitoring fault resolution progress and escalation procedures for the fault tickets that are reaching and have already exceeded the fault-resolution SLA.
- Monitoring of network performance. This should include the definition of: network performance data; analysis to be performed on the data given; the meaning of the analysis and any proactive/preventive network maintenance to be performed as a result of the analysis.

The example in Section 9.5.1.1 illustrates how all the above can be used as part of the network monitoring process. The network monitoring processes vary depending on the service requirements and the network element used. The example should be of a high enough level to provide a basic guide to develop the network management process for the service concerned.

9.5.1.1 Example High-Level Network Monitoring Process

A high-level network monitoring processes is illustrated in Figure 9.5 and a description of the process is given in Table 9.5.

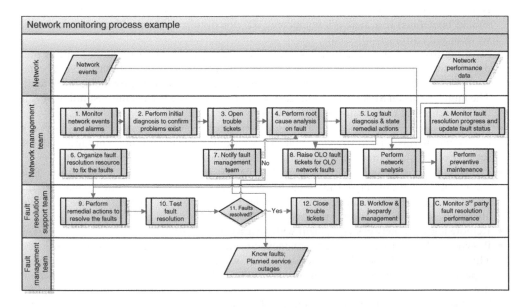

Figure 9.5   Example high-level network monitoring process.

Table 9.5   Description of a network monitoring process.

Steps	Actions/tasks description
Network management team	
1	Monitoring network events and alarms on the network management system.
2	When network alarms arise, perform initial investigation and use diagnostic tools where appropriate. Prioritize network problems in accordance with alarm severity.
3	If network alarms prove to be genuine problems after initial investigation, then trouble tickets are raised in the trouble ticketing/fault management system. As part of raising trouble tickets, fault severity categories are assigned for each trouble ticket.
4	Perform root cause analysis. This may include the correlation of alarms and events, remote logging onto the network element to view event logs, performing remote testing on network interface and so on.
5	The fault diagnosis should be logged and actions to remedy the situation should be stated. All the diagnostic tests, test results and reasoning for diagnostics should also be logged as part of the fault resolution on the trouble ticket. The trouble ticket status should be updated to reflect the state of play.
6	Once the fault resolution actions are known, resources required to fix the faults are identified and organized.
7	When the severity and impact of the faults are known, the information is passed to the fault management team to avoid unnecessary fault tickets being raised by the customers/end users.
8	If the faults lie with OLO network equipment, fault tickets with OLO are raised.
A	Work flow management of fault resolution progression should be monitored. Internal SLA trigger points at various steps of the end-to-end process should be set up to ensure the fault SLA is not jeopardized. If the trouble tickets are about to go into jeopardy, the escalation process should be invoked. Monitoring of fault resolution progress includes obtaining estimated time to resolve faults from third-party support agencies and OLO.
D	Taking the network events and network performance data as input, network performance analysis is performed.
E	Preventive maintenance may be required resulting from the analysis of the network performance data.
Fault resolution support team	
9	Carry out fault resolution actions as specified from the network management team.
10	Test the network and service after fault resolution actions have been carried out to ensure the faults have been resolved.
11	Check if the faults have been resolved. If they have not, further diagnostics will be required.
12	Close fault tickets when faults have been resolved.
B	Work flow management of fault resolution progress should be monitored to ensure fault ticket SLAs are not jeopardized. If the fault tickets are about to go into jeopardy, alarms should be raised to the management for more attention.
C	The performance of the fault resolution agencies is monitored to ensure the SLAs with third-party fault resolution agencies (e.g. OLO fault resolution team or subcontractors for field engineering work) are not exceeded.

9.5.2 Network Performance Management

Network performance management processes are designed to help operational personnel to keep the network operating in an optimum state by performing network data analysis and taking preemptive actions before network faults occur. For example, one of the items for network performance management could be the monitoring of CPU or memory utilization of network equipment. Unusually high CPU utilization under normal load conditions may indicate a potential fault with one of the processors within the network element. Early investigation and preventive actions may avoid the development of fault scenarios and, in turn, prevent the occurrence of service degradation.

The network performance management processes should contain (but are not limited to) procedures/guidelines for:

- setting up and using network performance monitoring tools;
- defining the parameter to be monitored;
- setting the thresholds for parameters to be monitored;
- defining the reports and reporting formats for the required analysis;
- setting up the network performance tool to perform the required analysis on the network parameters;
- interpreting the analysis results or trends and actions to be taken from the analysis results.

Network performance management processes are specific to the network element, the technology used and the services' performance requirements (i.e. the service performance SLAs). Therefore, the network performance management procedures should be defined in conjunction with service performance SLAs and the KPIs for the service.

9.5.3 Network Disaster Recovery

The network disaster recovery process is invoked when major network events occur. This could be a node going down, fire at one of the switch sites, a major transmission point going down, fiber break between major transmission points in the network and so on. Disaster recovery processes are dependent on the network technology that supports the service and should cover all disaster scenarios one can think of. The impact on the service/services should be assessed for each disaster scenario. The disaster recovery processes are technology dependent rather than service dependent. However, network disaster recovery processes should form part of the company's business continuity processes for the entire operation of the company, rather than just based on a certain piece of network equipment/technology or based on a particular service. This is because certain scenarios (e.g. fire at major switch sites) may involve the relocation of people and reconfiguring the network to divert network traffic out of the area. With good network resiliency design, the service impact during network disaster scenarios can be minimized.

9.5.4 Network Configuration Management

Configuration of the network should be under strict control. Network configuration management processes are defined to minimize unnecessary network and service outage due to errors introduced from changes in the network configuration. Network configuration management procedures should control what is in the network, log any changes made to the network and be closely linked with any network security procedures defined in the company. As part of the process:

- The authorization for network configuration changes should be defined.
- The amount of testing to be carried out on new a network configuration before implementation should also be stated. The objective of testing is to minimize the risk and impact on the current network and services being operated. This is especially important for software upgrades on network elements.
- Guidelines on when (time of day or months of the year, depending on traffic pattern) network configuration changes can take place.
- The procedures to follow (e.g. backup required data and network element shutdown procedures) when making network configuration changes should be described.
- Any service-/technology-specific guidelines that are required should be explained.
- A procedure on the use of any network configuration management tools should also be illustrated.

9.5.5 Network Maintenance

Network maintenance procedures include:

- preventive maintenance actions to be carried out as identified as part of the network monitoring analysis activities (as stated in Section 9.5.1);
- methods and analysis to recognize problem trends;
- preventive actions for potential problems;
- routine maintenance and archiving of network data (e.g. historical data of network problems and their resolution, historical network performance data).

Maintenance procedures as advised by the network element supplier should also be followed. Detailed work instructions for routine maintenance should include things like:

- routine maintenance tasks to be carried out;
- precautions to be taken before each routine maintenance tasks are to be performed;
- backup of configuration and data;
- time intervals for each maintenance task;
- security implications.

Network maintenance procedures are very specific to the network element being used and the load of the network equipment. Regular maintenance is to be performed to ensure the network is operating in an effective manner. Care is taken to ensure the maintenance work required for protecting the validity of the warranty given by the equipment supplier and/or any support contract arrangements is carried out. The network maintenance processes should also include monitoring of the performance of any maintenance suppliers to ensure that all the routine maintenance tasks are performed at the specified intervals and are of good standard, and that the supplier's maintenance contracts are fulfilled.

9.6 Network Traffic Management, Network Capacity Management and Network Planning Processes

The network traffic management, network capacity management and network planning processes are closely related, where their objectives are to maximize the network asset utilization and, at the same time, provide good service quality levels. Network traffic monitoring and management processes ensure the network is used in the most efficient manner and provide the actual network usage data as input into the network capacity management process. Network capacity management processes take the input from the traffic management process, the forecast information from product management, the potential capacity demand from the sales engagement process and the actual capacity demand from the network provisioning process to formulate the potential network capacity requirements for the service. These network capacity requirements are then fed into the network planning and build process, where additional capacity requirements are translated into network design and network build. For the service being designed, we need to ensure the high-level processes for the three areas are defined and linked together.

9.6.1 Traffic Monitoring and Management Processes

Traffic monitoring and management processes are defined to ensure the traffic levels within the network are managed properly and that the network resources are used efficiently. The processes that help to achieve this may included (but are not limited to):

- processes that define the traffic parameters to monitor, the threshold levels to set and the actions to perform when thresholds have been exceeded;
- detailed work instructions to analyze network traffic patterns, identify network congestion scenarios and detect traffic hot spots using the traffic analysis and management tools available.
- detailed work instructions to divert network traffic in accordance with traffic type when the network is congested due to unexpected demands or network events;
- detailed instructions for actions to be taken when network traffic hot spots have been identified;
- detailed work instructions to interpret traffic analysis reports/results and actions required (e.g. change in traffic shaping rules or change in traffic routing/cost of traffic routes) from the analysis results with predefined conditions;
- detailed instructions to define traffic analysis requirements and further analysis reports using the traffic management tools being used;
- detailed instructions to simulate potential traffic scenarios before changing traffic management rules using the traffic management tools.

9.6.2 Capacity Management Process

Before the capacity management process can be defined, one needs to determine the network capacity management policy/strategy for the service. Will the network capacity be provided 'just in time' to meet the demand, or will it be built when the customers place orders, or do you pre-build network capacity according to network trend and forecasts? Different network capacity management policies can be applied to different parts of the network supporting the service. For example, one may choose to have a just-in-time capacity management policy for the access network (because of shorter lead times), while the core network capacity for the service may be provided on a pre-build basis (because of long lead times and cost effectiveness). Therefore, the network capacity management policy is dependent on the cost and lead time required for building additional network capacity. Customers'/end-users' service and network provisioning SLAs and the amount of capacity spend available for the financial year are also influencing factors for the capacity management policy.

Capacity management processes for the service may include:

- network capacity utilization guidelines for the service, including the network utilization levels and parameters within which the network equipment should be operating for the services;
- a process to formulate network capacity requirements as input into the network planning process by taking into account the potential customer demand from the sales engagement process, the actual demand from the network provisioning process and the forecast from product management;
- procedures to define capacity management reports for the capacity management system and tools being used;
- guidelines for interpreting the analysis/reports from the capacity management system to identify potential capacity shortfall and actions required;
- process/guidelines to perform trend analysis related to capacity uptake for the service;
- detailed work instructions for the use of tools to generate network demand scenarios from the forecast and to produce trends analysis to assess the potential capacity requirements;
- detailed work instructions for the use of the traffic management tool to model potential network behavior in the scenario where the network capacity is not built within the timeframe against the estimated potential network capacity requirements, taking into account tolerance levels of the network.

9.6.3 Network Planning and Build Process

Network planning processes define the activities that need to be undertaken for formulating network expansion plans, building the additional network capacity required to fulfill potential customer/end-user demand, and obtaining business approval for the network expansion plans. In general, the network planning and build processes are likely to be nonservice specific and are applied across all the services being operated in the network. However, there will be service-specific processes to cater for different service requirements (e.g. definition of network planning rules for the service).

General network planning and build processes may include:

- network design process for capacity expansion;
- network build funding/business case approval process;
- procedures for installation and testing of new network elements (both existing network technologies and new technologies);
- procedures for upgrading network elements;
- managing and tracking progress of the additional network capacity being built (i.e. project manage all network build activities), including the management of external suppliers for the network build activities;
- procedures to update network inventory and network management systems for new network elements introduced and network element upgrades;
- maintaining network inventory and tracking all network elements that are in the network.

9.7 System Support and Maintenance Process

System support and maintenance processes are similar to those of network management processes. They are mainly divided into two areas: corrective maintenance and preventive maintenance.

Corrective maintenance actions are unscheduled maintenance activities/actions to restore functions or the system after failures. Processes in the corrective maintenance areas should include (but are not limited to):

- system monitoring;
- system fault management;
- system disaster recovery processes.

Preventive maintenance activities are scheduled maintenance actions in accordance with the manufacturers' recommendations to ensure that the systems are running at their optimum levels. Process areas of preventive maintenance include (but are not limited to):

- system performance management;
- system maintenance (including monitoring the performance of third-party support agencies);
- system configuration management;
- system security procedures.

9.7.1 System Monitoring and Fault Management

Operational and business support systems play an important role (one as important as the network) in providing a good service. Without the support systems functioning, not many of the operational tasks can be performed, nor, for some services, can calls and session set-ups take place. The service operations or the service itself can grind to a halt. Therefore, the health of the systems is monitored and system alarms should be attended to to avoid service degradation. The systems monitoring and fault management processes should include detailed work instructions for:

- Opening and closing trouble tickets for the trouble ticketing system logging system faults. This should define the minimum set of information to be included (e.g. system equipment ID, description of the problem, severity levels, services and customers affected) as part of the fault ticket.
- Selecting trouble ticket severity categories. The category definition can be the same as that of the fault management process for customer-/end-user-reported faults. A fault resolution SLA for each category should also be stated. The fault resolution SLA is usually defined in conjunction with the SLAs.
- Problem diagnostics and troubleshooting for each of the support systems within the service. This should include a list of diagnostic tests to be applied for different symptoms or alarm types. Corrective actions and fault resolution agencies for each potential system problem should also be stated.
- Precautions to be taken during system problem diagnostics.
- Passing fault details to the service management team/process. This should include details like system fault severity levels, service and customer impacted and assessment of the extent of service degradation, estimated fault resolution times.
- Stating the meaning of the different trouble ticket status. For example, trouble ticket status can be: open – await investigation; problem diagnosed – await to be fix (stating timescale); closed.
- Tests to be performed for each of the fault conditions after fault resolution actions have been applied.
- The management of fault resolution agencies and resources. This could be internal resource (i.e. field engineers) or third-party support agencies (e.g. suppliers). The purpose is to ensure they perform the tasks/requests within the agreed SLA.
- Monitoring fault resolution progress and escalation procedures for the fault tickets that are reaching and those exceeding the fault-resolution SLA.

System monitoring and system fault management processes are very similar to those of the network monitoring processes. Please refer to Section 9.5.1 for a process example and adapt it for the systems and the service requirements in use.

9.7.2 System Disaster Recovery

System disaster recovery processes are activated during failures or when system malfunctions occur. Recovery procedures should include the recovery of:

- system applications;
- system files;
- system data;
- data of customer/end user that is on the system after system failures.

System resiliency should be part of the system design (e.g. dual processors, parallel disc access) to minimize the loss of service in a system failure scenario. A separate redundancy system in a different geographic location is also recommended to minimize the impact if a disaster should occur (e.g. total loss of power) at sites where the systems are located/hosted. A regular backup and offsite archive procedure should be defined as part of the system disaster recovery process. The system disaster recovery process should form part of the overall business continuity process of the company.

9.7.3 System Performance Management

System performance management processes ensure that all the system performance parameters are monitored and that any necessary preventive actions for system performance degradation are carried out accordingly. Process areas to consider for systems performance management include:

- setting the system parameters and their thresholds to be monitored on the system performance monitoring system;
- monitoring of system performance using the defined performance metric or reports;
- defining additional system performance reports and analysis;
- interpretation of the system performance analysis results or trends and preventive actions to be taken (if any).

Since many systems within the service may have a direct impact on the service or customers'/end-users' perceptions of the service performance, it is important to define the system performance parameters in conjunction with the service performance SLA or the KPI for the service.

9.7.4 System Capacity Monitoring and Management

System capacity monitoring and management processes should take into account the system performance management reports and service forecast requirements to formulate additional system capacity requirements. Implementations of the required additional system capacity are scheduled accordingly.

System capacity management processes for the service may include:

- Guidelines for system capacity utilization for each system operating the service, including the CPU utilization levels, memory usage levels, disk space and other relevant system operating parameter for the services.
- A process to formulate system capacity requirements from the system performance management and trend reports and service forecasts requirements.
- Procedures to define system capacity management reports for the capacity management system and tools being used (if any).
- Guidelines for interpreting the analysis/reports from the system capacity management system for identifying potential system capacity shortfall. Actions to be take for certain results/predefined conditions should also be defined.
- Process/guidelines to perform trend analysis related to system capacity uptake for the service.
- Guidelines for scheduling system upgrades for each system.
- Procedures (installation and testing) for performing hardware upgrades (for additional system capacity) for all the systems operating the service.

9.7.5 System Maintenance

System maintenance procedures are normally carried out as recommended by the system suppliers. System maintenance processes may include:

- system files, applications and data backup;
- archiving system files, applications and data to offsite locations;
- management of backup storage media, including access security to backup materials;
- disk space management, including the management of disk quotas, performing disk partitioning;

- system maintenance tasks to keep the warranty validity as advised by the system supplier;
- system maintenance upgrades procedures;
- management and monitoring the performance of third-party maintenance agencies to ensure all the necessary maintenance tasks are carried out successfully;
- other maintenance tasks as recommended by system suppliers.

9.7.6 System Configuration Management

System configuration management plays an important role in maintaining a stable service operational environment. The following minimum set of procedures should be defined as part of the system configuration management processes:

- The authorization (i.e. who is authorized) for system configuration changes.
- The amount of testing to be carried on new system configuration/software before implementation. The objective of the testing to be performed is to minimize risk and impact on the current systems and services being operated. This is especially important for software upgrades on the systems.
- Guidelines on when (time of day or months of the year, depending on traffic pattern) the system configuration changes can take place.
- The procedures to follow when making system configuration changes.
- The use of any system configuration management tools.
- Roll-back procedures for upgrade failures or configuration changes.

9.7.7 System Security Procedures

System security procedures should be part of the company's security policy rather than defined on a per service basis, unless special security requirements are imposed as part of the service. Industrial standards for security policies and procedures can be found in ISO/IEC 27001 [10].

9.8 Revenue Assurance Processes

Revenue assurance processes relate to collecting revenue from the end users/customers, as well as in preventing revenue leakage occurring for the service. These processes are probably not service specific, but one needs to ensure that the business-as-usual processes for billing and revenue assurance cover the service being designed. Areas to be considered include (but are not limited to):

- billing and billing fulfillment;
- customer/end-user payment collection and management;
- bill reconciliation;
- revenue assurance and assessment activities;
- fraud prevention.

9.8.1 Billing and Billing Fulfillment

Most of the billing functions and billing fulfillment should be performed by the billing system. However, operational processes for billing may be required for:

- Setting up customer billing accounts, service pricing and payment terms (if not automated) and so on.

- Dealing with exceptions generated by the billing system. For each billing exception, actions to be taken should be defined as part of the billing exceptions process.
- Billing fulfillment (especially for paper bills). The process defines where the billing data is to be sent and the distribution mechanism for bills.
- Monitoring any outsourced billing activities, ensuring that the outsourced agencies are performing as contracted.
- Performing random bill-accuracy tests.

It is not uncommon to have customized manual billing for bespoke customer solutions. Bespoke manual billing processes should be linked to the service and provisioning processes for the service and should include the authorization requirements for manual billing and how the payments of these manual bills are tracked and accounted for.

9.8.2 Customer/End-User Payment Collection and Management

Payment collection is probably the most important item in the revenue assurance area (probably the most important part of the service, too). Most of the payment collection mechanisms (e.g. direct debit, payment through banks) should be automated, but operational processes are required to carry out checks and deal with payment exceptions. After going through credits checks for the customers/end users, the risk of bad debtors should be minimized. However, this does not mean that bad debts will not exist. It is suggested that the following processes are to be developed for the service (if they do not already exist as part of the 'business as usual' processes). Additional processes may be required, depending on the set-up of the operational environment.

- Detailed instructions for handling payment collection for each payment method. For example, it should include detailed instructions for setting up direct debits for end users/customers and handling check payments and so on. These processes should also contain spot checks on payments made and handling any payment exceptions (e.g. authorization failure during direct debit set-up).
- A process and detailed instructions for authorization of service termination for customers/end users with bad debts. The process should include the criteria for service termination due to bad debts and the authorization required to terminate the service. This process should be linked to the customer/end-user service termination process stated in Section 9.3.3. Depending on the actions defined, one might want to consider service suspension before terminating the service. This gives the customers/end users the chance to pay and provide a more positive service experience. If that is the case, then additional processes and system development might be required in the service provisioning/deprovisioning area.
- A process to monitor and identify overdue payments and actions to be taken under defined scenarios (e.g. actions for payment overdue for 1 month and actions for payments that are overdue for 2 months).
- A process/guidelines for collecting debt that is over a certain amount and escalation process for handling customers/end users with bad debts. The process should define the point at which the service should be terminated and if any legal proceeding is required.

9.8.3 Bill Reconciliation

Billing from other network operators (OLOs) are mainly for roaming charges, network interconnections charges from exchanging network traffic and network capacity bought from OLOs to fulfill customer/end-user orders. Reconciling these bills with the internal records is one of the items that will help to minimize revenue leakage. Although most of the reconciliation will be performed by the systems (impossible to check manually due to volume of data), processes are required to perform spot checks and handle exceptions. From a service design point of view, you need to ensure that the items to be reconciled for the service are covered by the existing reconciliation system and that the current processes will cater for any additional reconciliation items for the service. Otherwise, the following processes will need to be developed:

- a process to input OLO billing data into the reconciliation system;
- a process to deal with exceptions arising from the additional reconciliation items for the service;
- a process to define or change the margin of error for OLO reconciliation exceptions on the reconciliation system;
- detailed work instructions to use the bill reconciliation system for the additional reconciliation items and to produce any additional reports.

9.8.4 Revenue Assurance and Assessment Activities

As explained in Chapter 8, revenue assurance and assessment activities are cross-functional activities, involving the collection of data from various departments and systems. Depending on the amount of revenue assurance activities to be performed, operational processes are required for each activity. Revenue assurance and assessment activities may include (not are not limited to):

- using revenue account reports to match with the actual revenue for the service;
- using the service revenue reports to check against total revenue and actual revenue collected;
- monitoring CDR volume and the total number of call minutes against what is being billed.
- checking for unauthorized service credits to customers/end users.

Processes to obtain and use the data will be required. Revenue assurance and assessment activities may or may not be service specific. From a Service Design perspective, one needs to ensure that the service being designed have taken into account of the revenue assurance requirements and that the data and reports required are available for such purposes.

9.8.5 Fraud Prevention

Fraud detection is one of the means to prevent revenue leakage. Fraud prevention processes are designed for fraud detection and to minimize fraudulent activities. These can be on a per service basis or company-wide processes. When designing the service, you need to ensure that fraudulent opportunities are minimized and that operational processes are in place to identify potential fraudulent activities for the service. In some cases, there will be requirements for fraud detection systems to assist with such tasks. Detailed work instructions for the use of such a system will be needed. Please see Chapter 8 for the system functions for fraud prevention.

9.9 Process Mappings to eTOM Model

For readers who are familiar with the eTOM model [31], Table 9.6 shows the mapping between the process areas discussed in this chapter and those within the eTOM model.

Table 9.6 Mapping between the process areas discussed in this chapter and those within eTOM.

Processes discussed in this book	eTOM processes (process identifier)
9.1 Sales engagement process	Market fulfillment response (1.1.1.3)
	Selling (1.1.1.4)
	Support customer interface management (1.1.1.1.1)
	Support market fulfillment (1.1.1.1.6)
	Support selling (1.1.1.1.7)
	Manage customer inventory (1.1.1.1.10)
	Manage sales inventory (1.1.1.1.12)
	Design solution (1.1.2.2.1)
9.2 Customer service process	
9.2.1 Service enquiries process	Selling (1.1.1.4)
9.2.2 Complaint handling	Bill inquiry handling (1.1.1.12)
9.2.3 Billing enquiries	Bill inquiry handling (1.1.1.12)
	Support bill inquiry handling (1.1.1.1.15)
9.2.4 Technical support	
9.2.5 Order management	Order handling (1.1.1.5)
	Support order handling (1.1.1.1.2)
	Issue service orders (1.1.2.2.7)
9.2.6 Service change requests: MACs	Order handling (1.1.1.5)
9.2.7 Service and network termination, cancelation and customer retention	Recovery service (1.1.2.2.10)
	Retention and loyalty (1.1.1.9)
	Support of retention and loyalty (1.1.1.1.5)
9.2.8 Customer/end-user migration	Manage product exit (1.2.1.5.8)
	Manage service exit (1.2.2.3.7)
	Manage resource exit (1.2.3.3.7)
9.2.9 Customer fault management	Problem handling (1.1.1.6)
	Support problem handling (1.1.1.1.3)
9.3 Service and network provisioning	
9.3.1 End-user service and network provisioning	Service configuration & activation (1.1.2.2)
9.3.2 Customer service and network provisioning	
9.3.3 Service and network termination/cancelation	
9.4 Service management processes	Customer QoS/SLA management (1.1.1.7)
	Support customer QoS/SLA (1.1.1.1.8)
	Enable service quality management (1.1.2.1.4)
	Support service & specific instance rating (1.1.2.1.5)
	Service quality management (1.1.2.4)
	Resource data collection & distribution (1.1.3.5)
9.5 Network management and maintenance processes	
9.5.1 Network monitoring	Service problem management (1.1.2.3)
9.5.2 Network performance management	Support service problem management (1.1.2.1.3)
9.5.3 Network disaster recovery	Resource management &operations (RM&O)
9.5.4 Network configuration management	Support and readiness (1.1.3.1)
9.5.5 Network maintenance	Resource performance management (1.1.3.4)
	Resource data collection & distribution (1.1.3.5)
9.6 Network capacity and traffic management and network planning processes	
9.6.1 Traffic monitoring and management processes	Enable service configuration & activation (1.1.2.1.2)

(*continued overleaf*)

Table 9.6   (*continued*).

Processes discussed in this book	eTOM processes (process identifier)
9.6.2 Capacity management process	RM&O support and readiness (1.1.3.1)
9.6.3 Network planning and build process	Resource provisioning (1.1.3.2)
9.7 System support and maintenance process	
9.7.1 System monitoring and fault management	
9.7.2 System disaster recovery	Resource trouble management (1.1.3.3)
9.7.3 System performance management	RM&O support and readiness (1.1.3.1)
9.7.4 System capacity monitoring and management	Resource performance management (1.1.3.4)
9.7.5 System maintenance	Resource data collection & distribution (1.1.3.5)
9.7.6 System configuration management	Resource provisioning (1.1.3.2)
9.7.7 System security procedures	
9.8 Revenue assurance	
9.8.1 Billing and billing fulfillment	Bill invoice management (1.1.1.10)
	Support bill invoice management (1.1.1.1.13)
9.8.2 Customer/end-user payment collection and management	Bill payment and receivable (1.1.1.11)
	Billing and collection management (1.1.1.8)
	Support bill payment and receivable (1.1.1.1.14)
9.8.3 Bill reconciliation	S/P settlements and payments management (1.1.4.5)
9.8.4 Revenue assurance and assessment activities	Service & specific instance rating (1.1.2.5)

10

Implementation Strategy

With all the design activities defined, it is time to think about implementation and implementation strategy. This raises a number of challenges. How are you going to make it happen? What approach should you take? Do you do everything in-house or do you outsource part of the design and implementation? Do you have sufficient resources in-house to do everything? Will outsourcing give you a quicker delivery? Do you have all the skills required? What are the cost and risks involved if you outsource your development against an in-house development team?

This chapter outlines ways to define the implementation strategy and factors to consider when defining the strategy. It also covers the various implementation scenarios readers may encounter. The benefits and risks of each scenario will be highlighted and ways to mitigate those risks are also recommended.

10.1 What is Implementation?

Implementation of the project is about making the project real and making it happen. Things to consider at the start of project implementation include:

- what needs to be done;
- resource requirements for each activity;
- timescale of each project item;
- production of project plans for the different activities with interactivity dependencies.

Figure 10.1 illustrates the seven steps for implementation.

10.1.1 Planning

The implementation planning stage starts when the activities to be performed or items to be implemented have been/are being defined. In some projects, it starts when the project objectives have been defined. We should think about what needs to be done to achieve the objectives and how to go about doing it. Quantifying what needs to be delivered is probably the first thing that needs to happen at this stage. For service design activities, what needs to be done should be detailed in the

---

Successful Service Design for Telecommunications   Sauming Pang
© 2009 John Wiley & Sons, Ltd

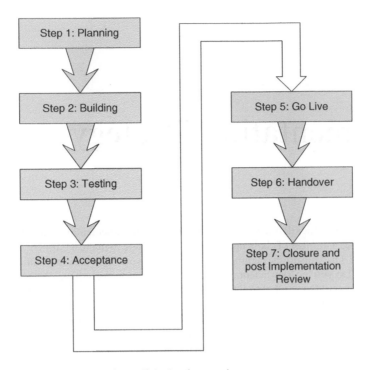

Figure 10.1   Implementation steps.

technical service solution document and various high-level designs produced, as discussed in Chapters 4 and 5.

During the planning stage, the implementation strategy (please see Section 10.2 for details) and the high-level implementation plan are defined. The implementation strategy defines the approach to the project, whilst the high-level implementation plan illustrates the high-level tasks to be performed, the interdependencies between the tasks, resource requirements and ownership of the activities. The project plan is also an output from the planning stage, where all the detail activities to be performed are scheduled and resources are allocated to each of the activities.

10.1.2 Building

This is when the implementation activities take place. For the service implementation, this is when the systems and network are implemented and built as per design. For system developments, this will involve developing, installing and configuring the systems as per the system designs for each of the systems that is within the scope of the service. From a network development point of view, this is the stage where network equipment is delivered, installed and configured as specified in the detail network design document. From an operational point of view, this is when system users and operational personnel are trained to use the operational support systems being implemented and when they are trained on the operational process for the service. If additional operational personnel are required, then recruitment and desk space allocation, etc. will take place during the building stage of the implementation process.

10.1.3 Testing

The activities within the testing stage ensure what has been built works as designed and that they fulfill the requirements specified. For system developments, system testing activities are performed to ensure the systems work as designed. For the network development, the testing activities ensure that the network is built as per design and that the configuration of the network equipment is performing the network features as expected. In addition, the testing activities also make certain that the new network elements do not jeopardize the current stable network operational environment. From the operational point of view, operational testing ensures that the operational processes defined are fit for purpose and that all operational personnel have been trained and are ready to operate the service. Further description on the different types and stages of testing can be found in Chapter 11.

10.1.4 Acceptance

After all the testing activities have been carried out, operational personnel need to accept the management responsibilities of the systems and the network elements for the service. From the systems point of view, the system users need to accept that the systems are performing functions as specified and that they are fit for purpose. Normally, system users will carry out user acceptance testing before accepting the systems. The system operational team also needs to accept the management responsibilities of the systems. They should have received training on the systems to be managed.

From a network development point of view, the network operational teams need to be satisfied that the new network elements are stable and are working as designed. At the acceptance stage, the teams should have been trained to operate and manage the new network elements. They should be ready to accept the responsibilities for the new network elements. If the solution is built by third-party suppliers, then this is where the system/network solution is signed off and accepted and where suppliers' contractual commitments are reviewed. Upon successful completion of acceptance testing, payments to the third parties are made.

10.1.5 Go-Live

Go-live is when live customer/end-user traffic is taken onto the service. Depending on the implementation strategy and service launch strategy, go-live may also mean having some trial customers/end users to try out the service. It is always a good idea to run a trial before launching the service with a high volume of customers/end users. During the early stages of service launch, the design and implementation team should be kept in place to monitor the service performance and to resolve any teething problems the service might have.

10.1.6 Handover

Handover is when the project team is dissolved and all the support of the service is handed over to the operational teams.

10.1.7 Closure and Post-Implementation Review

After handing over the service to the operational teams, the project can be closed. Any budget and financial matters are settled and all the project documentation is completed.

It is also a good idea to hold a post-implementation review to assess:

- How did the project go?
- Have the project objectives been met?
- What has gone well?
- What can be done better?
- What are learning points of the project?
- Are the stakeholders of the project satisfied with the implementation outcome?

A post-service launch review should be held, typically after 3 months of service going live. The purpose of the review is to ensure that the service is operating smoothly and that no outstanding issues remain. Please see Chapter 11 for further details.

10.2 What is an Implementation Strategy?

An implementation strategy defines the approach by which the project is to be executed. It states why you are doing the project, the end goal and the objective of the project (the what) and steps (the how) to achieve that goal. The strategy should take you from the start of project implementation through to the end of implementation and project closure. The strategy is defined at the start of the project or at the start of the implementation phase of the project to obtain buy-in of all the stakeholders to the project and to secure the resources required and sometimes to secure the funding for project implementation. As part of the exercise, you are also obtaining the buy-in for the high-level delivery milestones and timescales of the project implementation.

Within the strategy, you might also want to define the exit and entry criteria for all the steps and stages listed in Section 10.1. This will ensure that everything that ought to have happened has been done. You might also want to define which stakeholders within the project can sign off each implementation step or stage.

Stating the roles and responsibilities for all stakeholders and parties at all the different stages is also part of the strategy. This enables the whole project team to understand clearly the ownership of various implementation activities and the accountability and demarcation points of the various stakeholders within the implementation team. This also helps with the transparency of the project for other parties involved in and outside the implementation project. Resource requirements can also form part of the roles and responsibilities definition. As part of the roles and responsibilities definition, it is useful to define the implementation team structure. The escalation path for issues and unforeseen events should also be defined within the implementation strategy.

Phasing of the project and what is to be included in each phase is one of the key elements within the implementation strategy. In a large project, it is not unusual to implement the solution in several phases. This is normally due to timescale or resource constraints.

10.3 Why Do We Need an Implementation Strategy?

You might ask: why do we need an implementation strategy? Does the implementation project plan not define what needs to be done and who does it? It is true that the implementation project plan defines the tasks that need to be done, the timescale and resources to perform these tasks and the interdependencies between the various tasks. However, it does not cover the high-level approach to the project (or problem), the phasing of the project or how the project is to be implemented. The implementation project plan is a result of the implementation strategy defining the details of task to be performed and their interdependencies, not the approach and how to implement the project. The implementation project should not dictate the implementation strategy; rather, it should be the reverse, namely that the implementation strategy should set the direction for the implementation project plan.

One might also argue that the objective of the project, the structure of the project and so on are stated in the project initiation document (PID). This is correct. However, the PID does not fully define the approach to the implementation. Some do use the PID to set the direction of the project and use it as a documentation vehicle for the project implementation strategy.

It is often useful to define the approach and strategy to the project so that everyone 'sings from the same hymn sheet' when implementing the project. This is especially important for large-scale projects with many stakeholders and implementation phases.

10.4 What Are the Steps and Approach to Take When Defining an Implementation Strategy?

When defining the strategy, one needs to know:

- What are the objectives? What are you trying to achieve and why you are doing it?
- How to achieve it?
- Who does it?

When you have answers to all the above questions, then you can start defining the implementation strategy and the implementation project plan.

10.4.1 Project Objectives: What are You Trying to Achieve

Project objectives define what you are trying to achieve and the end goal of the project. This also explains why you are doing the project. The project objectives could be:

- implementing a service (i.e. the solution fulfilling the service requirements) within a certain timeframe and of a budget of £x;
- satisfying certain quality objectives;
- achieving a certain level of customer satisfaction;
- fulfilling certain business objects;
- creating or fulfilling certain business benefits.

It is important that the objectives can be quantified and measured. When defining the project objectives, one should contain elements of team sprit or team communication as part of the project objectives. This can be fundamental to the success of the project, as the project team is made up of people and you need them to implement the defined tasks.

One also needs to consider the constraints the project might have. Some of the objectives, like time and budget, can also become the constraints of the project. The other constraints can be resource availability, geographic location and time zone differences.

10.4.2 How to Achieve it?

The actual items to be implemented (i.e. what needs to be done) should be stated in the various design documents, as covered in previous chapters. However, defining how and the sequence of events to make it happen lies within the implementation strategy. One might need to ask:

- Is there a logical sequent of events?
- What is the most efficient way to do it?
- How much parallel tasking is practically possible with the resource (people and platform availability) constraints?
- What are the interdependencies between the activities?

- What would happen if a certain activity is to be done in a certain way?
- How much risk is there if you go down route B rather than route A? Will taking route B still meet the project objectives?
- Will splitting the implementation into several phases help with time constraints or minimize project risks?
- What are the risks if the implementation is split into several phases?

In general terms, parallel tasking is the most efficient way to complete tasks in terms of timescale. However, sometimes it is not practical or it is impossible due to people's availability (e.g. you may be using the same set of resources for both activities and it is not possible to work on two things at the same time). Interdependencies between activities also dictate the sequence of events. Some activities are just sequential, as it is not possible or it is too high a risk to start before a certain task is completed. For example, if the network design is not complete, then it is not possible to start building the network solution.

Implementation can be carried out as a single, one-off event. Splitting the project into different phases is not unusual, especially if the project is on a very strict timescale or if delivering everything within the scope of the project is not practical. Other reasons for splitting the implementation of the project into different phases may include the risk being too high to implement everything at the same time (especially for large-scale projects) or the opportunity to roll back if something goes wrong, which may not exist if everything is to be implemented in one single phase. If the project is to be delivered in several phases, then the scoping of each phase needs to be clear and the project sponsor or the internal/external customers and the stakeholders within the project need to agree what is to be delivered in each phase.

10.4.3 Who Does it?

Who does it depends on:

- the skills available;
- the number of people required;
- the availability of the right resource;
- timescale required;
- cost of labor.

The following sources of resources can be used to perform the tasks:

- In-house – identify people within the company that have the skills and are available to perform the tasks.
- Outsource – identify a resource supplier (e.g. supplier, system integrator or consultancy firm) to provide the resource required for implementation.
- Contract – using contractors for a short period of time for individual tasks may also be a solution
- Combination of all the above – within a large-scale project, it is not unusual to do some of the tasks in-house and outsource some of the activities. The combination of a few contractors and in-house resources can also work well.

When outsourcing development activities, you need to ask whether you are outsourcing all your development or whether you only outsource a certain part of the development. Within each development area, do you outsource both the design and implementation, just the design or just the implementation?

The decision on outsourcing also depends on whether you require the skills to be retained in-house. If the skill required is only a one-off, then outsourcing or issuing a contract for someone to perform the tasks is viable. If the skill is required after the service is launched, then it might be wise and may be cheaper to recruit someone permanently to perform the role. It is not unusual to employ contractors for roles where specialist expertise is required or where there are skill shortages in the market and good permanent staff proves impossible to recruit.

Cost is, of course, a major consideration. In some instances it might be cheaper to outsource a development than to hire individual contractors or recruit permanent staff, where internal resources are not available. This is especially true for some design agencies and system integration companies who specialize in developing solutions for the technologies to be used. Instances where contractors might be an advantage over using outsourced agencies are: when the piece of work is relatively small, where the tasks are only to performed once or if the skill required is very low level (e.g. data entry), where it is cheaper to employ individuals to perform the tasks.

Other factors for consideration are the ownership of the tasks and who will be managing these resources and monitoring their progress. This is especially important if you outsource or use contractors to perform the tasks. It is preferable to have someone in-house to monitor, review and approve work that has been outsourced or contracted. This will minimize the risks of the project.

The criteria for choosing which strategy and how the project is achieved are very much dependent on the project objectives, what you are trying to achieve, the cost, resources and time constraints.

10.5 Implementation Strategy Example

The first step for forming the implementation strategy will be defining what you are trying to achieve and what you are implementing. To illustrate how an implementation strategy can be derived, I have used an example solution as described in Figure 10.2.

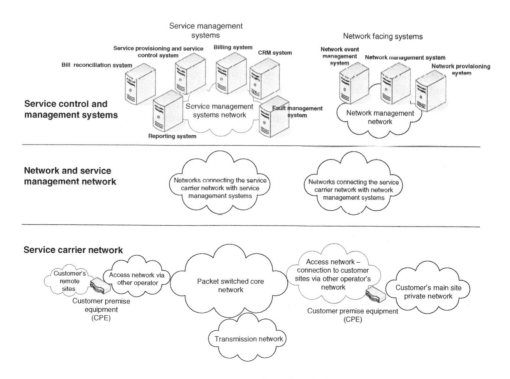

Figure 10.2 Example service solution.

In the example solution, the service we are trying to implement is a managed IP-VPN service with the service demarcation points being the CPE at a customer's remote sites and the CPE (service termination device) at the customer's main site. The network technology being implemented has been in the market for a few years and proves to be stable. Several new systems and system upgrades/developments are required. New operational processes are to be designed and implemented. Testing activities for this example service solution are covered in more detail in Chapter 11.

I have used generic terms of the network and systems rather than naming specific network devices and systems in order that the readers can adapt the principles easily into their service solutions. The tasks/activities are still relevant irrespective of the systems or network elements being used from an implementation perspective. The actual designs of the system, network and operational processes/detail instructions are, of course, specific to the network elements, the systems and the operational organization. However, the implementation activities can be used to form a conceptual view of the implementation strategy.

10.5.1 What are You Trying to Achieve?

In this example, the project objectives are to:

- implement the service solution within a budget of £2 750 000;
- implement on a timescale of 9 months;
- satisfy the trial customer that the service and the right level of support are provided;
- keep everyone in the project team informed of the progress, issues and risks;
- minimize hierarchical escalation and decision making.

10.5.2 What Needs to be Done?

The network components to be implemented include:

- four new core network packet switches;
- additional transmission network for the core switches;
- design configuration of new CPE;
- access network to customer's main sites via other operators (OLOs);
- access network to customer's three remote sites via other operators, including the CPE at the trial customer's remote sites;
- installation of CPE/service termination device for a trial customer;
- network connecting the network management systems;
- network connecting the service management systems;
- network management system network;
- service management system network.

The system implementation activities within this example service solution include:

- design and implement a new service provisioning and service control system – the functionalities of this system also include the provisioning of all the network elements for the customers for the service;
- enhancements to the existing CRM/order management system;
- enhancements to the current billing system;
- enhancements to the bill reconciliation system;
- design and implement a new customer and KPI reporting system;
- enhancements to the existing fault management system;
- network provisioning system upgrade to cater for the new core network technologies;

- network element management system upgrade to cater for the new network technologies (core network packet switches and CPE);
- network management system upgrade to cater for the new network technologies and the additional features for the service.

The operational processes to be designed and implemented for this service include:

- sales engagement;
- customer service – product enquiries;
- service and network provisioning;
- bill enquiry;
- fault management;
- network management and maintenance;
- service management;
- system maintenance.

10.5.3 How to Achieve it?

In this example, the project is to be implemented all in one single phase. The benefit of implementing the project in phases is only marginal, as this service's features cannot function properly with a phased implementation.

10.5.4 General Approach

Network implementation. The core network and the CPE development are carried out in series due to resource constraints. The transmission network cannot be finalized until the core network design is complete. The rest of the network development's activities and the service management systems are carried out in parallel.

System and process implementation. All the system development and operational processes developments are implemented in parallel. Details of the sequence of activities to implement this project are described in Figure 10.3.

Integration and testing. The integration and testing activities for the solution are done in their logical groupings. Details of the integration and testing strategy for this service solution example are discussed in Chapter 11.

10.5.5 What if This Example Solution is to be Implemented in Phases?

If the project were to be split into phases to minimize risk, then the service could be implemented in two phases. Phase 1 includes the implementation of the:

- network development activities (all);
- new system development for service provisioning and service control system;
- enhancements to the CRM/order management system;
- enhancements to the billing system;
- network provisioning system upgrade to cater for the new core network technologies;
- network element management system upgrade to cater for the new network technologies;
- network management system upgrade to cater for the new network technologies and the features for the service;
- service and network provisioning process;
- network management and maintenance process;
- service management process;
- a trial customer using the service with dedicated support teams.

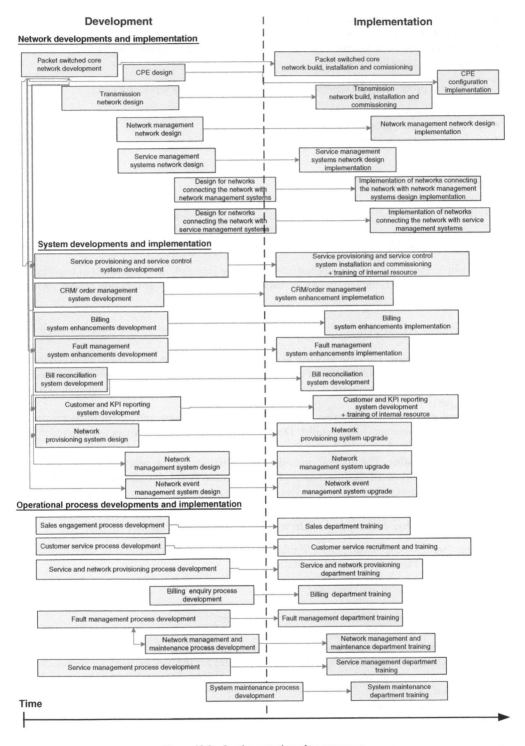

Figure 10.3 Implementation plan summary.

Phase 2 includes the implementation of the:

- enhancements to the bill reconciliation system;
- new customer and KPI reporting system;
- enhancements to the fault management system;
- sales engagement process;
- customer service – product enquiries process;
- bill enquiry process;
- fault management process;
- system maintenance process;
- trial customers using the full service solution.

The activities in phase 1 provide the minimum set of systems and network required to run the service on a very small scale with a small but highly skilled support team. This is not sustainable with more than 10 customers.

10.5.6 Who is Doing the Work? Who are the Stakeholders?

As stated in the previous section, who does the work very much depends on the skill availability of resources and the cost of labor. In this example, the design and implementation of the new network and service provisioning system and the new customer and KPI reporting system have been outsourced to the suppliers of the respective systems. This is due to the lack of internal skills and resource. The approvals for the systems are to be obtained from the internal owners of each system owner; that is, the head of provisioning systems and the head of reporting systems.

The stakeholders for each implementation activity (in program management terms – work package) are usually the owners of the functional areas concerned. For example, for the transmission network implementation, it can be the head of the transmission network or the head of the network department; and for the billing enhancements, it can the head of the billing department or the head of the IT/systems department. The stakeholders are dependent on the organization of the company, but they should have the responsibilities for designing or operating the network/system/ operational areas. Table 10.1 gives a summary of the different work packages, the resources that will be performing the tasks and the stakeholders for each work package.

Table 10.1 Example resource and stakeholders summary.

Implementation work packages	Who does the work	Stakeholder
Network development and implementation		
Core network packet switches	Core packet network design team	Head of customer network department
Transmission network	Transmission network design team	
CPE	Core packet network design team	
Network connecting network management systems	Internal data network design team	Head of internal network department
Network connecting service management systems	Internal data network design team	
Network management system network	Internal data network design team	
Service management system network	Internal data network design team	

(*continued overleaf*)

Table 10.1   (*continued*).

Implementation work packages	Who does the work	Stakeholder
System development and implementation		
Service provisioning and service control system	Outsourced to supplier with design approval from stakeholder	Head of service and network provisioning systems
Enhancement to the CRM/order management system	CRM system design team	Head of CRM systems
Enhancement to billing system	Billing system design team	Head of billing system
Bill reconciliation system enhancement	Bill reconciliation system design team	
Customer and KPI reporting system	Outsourced to supplier with design approval from stakeholder	Head of reporting systems
Fault management system enhancements	Fault management system design team	Head of fault management system
Network provisioning system upgrade	Network provisioning system design team	Head of network provisioning system
Network element management system upgrade	Network management system design team	Head of network management systems
Network management system upgrade		
Operational processes development and implementation		
Sales engagement process	Product manager and sales team	Head of sales
Service management process	Service management team	Head of customer services
Customer service process	Customer service team	
Fault management process	Fault management team	
Service and network provisioning process	Service and network provisioning team	Head of service provisioning
Bill enquiry process	Billing team	Head of billing
Network management and maintenance process	Network management team	Head of network operations
System maintenance process	System management and operations team	Head of system operations

10.5.7 Network Owners

In this example, the core packet network and the CPE are designed by the core network team, whilst the design and implementation of the network connecting from the network elements to the internal systems and the network connecting between the internal systems (network management and service management systems) are done by the internal data network design team. All the network designs are approved by the network design authority of the service.

10.5.8 System Owners

On the systems side, the design and implementation of the new service provisioning and service control system and the new customer and KPI reporting systems have been outsourced to their respective suppliers. New service provisioning and reporting systems are required because the existing internal provisioning and reporting systems do not have the functionalities or the capacity to cope with the new service and because the enhancements to these systems are not economical.

Since there is no internal resource available to develop a new system, it has been decided that an external system is to be procured. After going through the supplier selection process, suppliers are chosen for the two systems. The design and implementation of the two new systems are done by the suppliers. All the design activities are supervised and approved by the system design authority for the service and the heads of the respective system areas. Training of internal resources to use and operate the new systems also needs to take place.

The network element and network management system enhancements are done by the same team: the network management system design team. The rest of the system enhancements are done by the respective design teams of the existing systems. All the system designs are to be approved by the system design authority of the service.

10.5.9 Operational Process Owners

The high-level operational processes are produced by the process design authority for the service with heavy involvement from the respective operational departments. Detailed work instructions for the different areas are produced by the representatives or the change managers of the operational departments. All high-level operational processes and detailed work instructions are to be approved by the heads of the respective operational areas.

For stakeholder and ownership, an RACI (responsible, approval, consulted and informed) chart can be produced detailing who is responsible for the tasks, who is to approve the work and who needs to be consulted and informed.

10.5.10 Example Implementation Plan

A summary of the implementation plan for this service solution example is as described in Figure 10.3. To increase readers' understanding of potential interdependencies between implementation tasks, the implementation plan example has included the design and development activities as well as the building and implementation activities. Where possible, the dependencies between activities are shown. Please note that the durations of all the activities shown are not absolute timescales.

10.5.11 Network Implementation

For network development and implementation, the core packet network design activity is carried out in series with the CPE design. This is due to the resource constraints. The same goes with the network management network design activity and the design for connecting the network elements with the network management system, as well as the service management systems network design and the design for network connecting the network elements with the service management systems. Hence, both sets of those tasks are staggered. All four sets of designs have dependencies on each other, as well as on the service carrier network elements.

In this example, the implementation timescale for the network area is dictated by the delivery lead time of network elements. The start time in the network implementation phase is influenced by the lead time of the core network element supplier, rather than by resource constraints. The CPE configuration implementation suffers from the same resource constraints as the design phase.

10.5.12 System Implementation

On the system development and implementation side, all the developments are done in parallel and start at the earliest possible time. The design of the network provisioning system, the service provisioning and service control system, the network element management system and the network management systems all have dependencies on the design of the core packet network design and CPE design. Therefore, these tasks may start before the two network designs are done, but they cannot finish before the core network and CPE configuration designs are finalized. Similarly, the network management system design and network element management design activities span across both the network management network design and design for connecting the network elements with the network management system.

Most of the system enhancements implementation start times (except for network provisioning, network element and network management system upgrades) are constrained by the development and release cycle of the existing systems. It is not unusual to have several system enhancements (either for process improvements purposes or new service introduction) to be grouped together into one version of software release. The start time for the network provisioning, network element management and network management system upgrades are specific to the service being developed and the implementation timescale is constrained by the availability of the version of the software required.

10.5.13 Operational Implementation

As mentioned in the previous sections of this book, the definitions of various operational processes are dependent on the design of the systems and vice versa. Hence, the development of most of the operational processes starts at the same time as their system development counterparts, with the exception of the billing enquiries process. As the service definition dictates the design of the billing format, charging/pricing structure and, hence, the system design, the billing enquiry process development can only start after the billing system design is completed. It is also worth noting that the fault management and network management and maintenance processes are closely related as they are both dealing with network faults. Establishing good communication contact points between both processes and teams will help to minimize unnecessary fault tickets being raised and faults tickets can be closed more effectively. The system maintenance procedures are normally the last to finish after all the designs of the systems have been completed.

The implementation of the operational process in this context requires the identification of operational personnel and running training sessions with them. The training sessions will include both the operational processes/detailed work instructions to follow and the use of the new system/ system enhancements. In this example, it has been identified that a new customer service team is required. Therefore, the training can only take place when all or some of the new recruits start work. The rest of the training activities are constrained by resource availabilities. One also needs to be aware that you do not want to train people too early before the start of operational service testing. From past experience, people tend to forget what they are supposed to do without real-life practice. Therefore, it is a good idea to have the training as close to operational service testing as possible and for the training to include hands-on experience of the system or network technologies involved. The availability of network elements and systems for training purposes can also be a potential issue. A separate training environment may be required. Depending on the network element and system testing environment used, some training, especially technical training, may be performed using the testing facilities.

The implementation example given above has covered some of the possible implementation scenarios. These scenarios can be applied to different network, system and operational processes and services. The actual implementation tasks/activities may not be exactly the same for all services, but the dependencies between the various tasks should still apply.

11

Service Integration and Service Launch

Service integration and service launch activities directly follow the implementation stage. The service integration and service launch strategy could be part of the overall implementation strategy, as the implementation is not complete until the service has been handed over to operations. I have split this into a separate chapter for clarity.

For most service introductions, there are three main streams of development: network technology, support systems and operational processes (as described in the service building blocks in Chapter 6). Within each of the main development streams, there are substreams (sometimes known as work packages) for each functional area. All of them require integration for harmonious operation. So, how do you pull everything together? How do you integrate various parts of your solution? What and how much testing do you need to ensure the solution is working as designed? What type of test environment do you need? Who will be performing the integration and service launch activities? How do you launch the service effectively? What are the pitfalls? These questions should be answered in the integration strategy (as described in Section 11.2).

This chapter provides a guide for launching new services – from service integration strategy through to service operation. Examples will be given to illustrate the service integration strategy. The approach to introducing a new service into an already busy, but stable, operational environment is also discussed.

11.1 Service Integration Model

Service integration has many similarities with system integration in many respects, where the objectives of the activity are to:

- confirm (or otherwise) that the solution developed fulfills the requirements;
- ensure all the systems – in this instance, that all aspects of the service integrate and work harmoniously together;
- minimize the risk as a result of introducing the system – in this instance, the new service into the operational environment.

Successful Service Design for Telecommunications   Sauming Pang
© 2009 John Wiley & Sons, Ltd

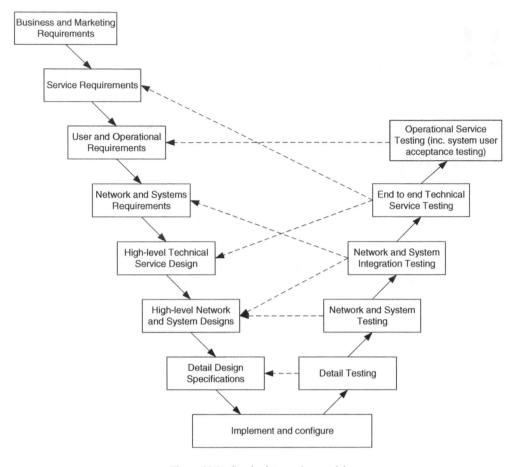

Figure 11.1   Service integration model.

Hence, the service integration model (Figure 11.1) is not dissimilar to that of the classic validation, verification and testing (VV&T) model (Figure 11.2) of software and system testing. Testing is the mechanism to confirm all the systems and network components within the service are well integrated together and that they performing the functions as required. This, in turn, lowers the risk in introducing a new service into the stable operational environment.

Like the principle of software development and the VV&T model, the verification and testing team for the service integration activity should be independent to (or different from) that of the design team. This way, the validation of the requirements and the verification of the design of the service will be impartial. The gaps in the requirements and the design can be identified more effectively.

Service integration activities are divided into five stages as illustrated in Figure 11.1:

- detail testing of individual network components and systems;
- network integration testing and system integration testing;
- network and system integration testing;
- end-to-end technical service testing;
- operational service testing.

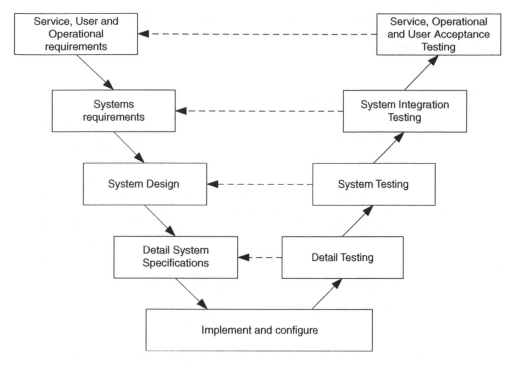

Figure 11.2   Software validation, verification and testing.

11.1.1 Detail Testing

Detail testing involves testing the individual network element and system against the respective detail design specification and requirements. For example, the service may use a range of CPE; the configurations on each piece of equipment are specified and tested to ensure the service features required function as expected.

From a system point of view, this is equivalent to the system testing within the VV&T cycle. All other lower level testing activities (e.g. unit testing) should have been completed successfully before reaching this stage.

This stage of testing ensures that the individual network elements and systems perform as specified and that they satisfy the network and system requirements for the service. If the individual system and network equipment does not pass the tests at this stage, then there is little chance that the service will be functioning as expected. Any test exceptions and requirements nonconformances should be logged and risk assessed before the next stage of integration activities starts.

11.1.2 Network Integration Testing

Network integrating testing activities mainly concentrate on testing the interoperability of all the network elements within the service network solution. This ensures that the various network elements within the solution are working together as designed and that the network requirements for the service are fulfilled. This interoperability testing should also confirm that no error occurs when passing network traffic between the various network elements within the service network solution. For example, the core network elements of a service will need to be integrated with the transmission network elements such that network traffic can be passed between various core network elements for the service.

11.1.3 System Integration Testing

After the individual system testing has been completed, systems that work together to form the solution for the service will need to be integrated. This is to confirm that all the individual systems can communicate together to form a systems solution and that no error exists (or no information is missing) when passing information between the systems. For example, all the business support systems will need to work together to form a customer-facing solution for the service; for example, the integration of the CRM system with the billing system so that all the necessary customer information is passed, without error, from the CRM system to the billing system. The system integration testing also verifies that the systems requirements and the high-level system designs have been fulfilled.

11.1.4 Network and System Integration Testing

Within the service solution, network-facing systems need to communicate with the network elements for the service to function and operate properly. Therefore, the integration of the network elements and the relevant systems is very important. Within a service, the network-facing systems normally include the network provisioning system, the network management system, the reporting systems and the billing/billing mediation system. For example, the network and system integration testing activities may include: provisioning end users onto the network using the network provisioning system; collation of network events from the network event management system from the network elements; collecting network performance data from the network elements for network performance reporting. The network and system integration testing also confirms that the network and systems requirements and the high-level network and system designs have been fulfilled.

11.1.5 End-to-End Technical Service Testing

End-to-end technical service testing verifies that the end-to-end service has been built and that the solution works together harmoniously. After integrating the network elements and the systems together, the technical service testing ensures that all the information flows work as expected and that no unexpected error has been produced. This stage of the integration confirms that the service is functioning harmoniously end to end, the service requirements have been met and that the high-level technical service design has been built. For example, one of the test scenarios can be the end-to-end provisioning for end users where end-user orders enter the end-user order management system and the order information is passed through to the network provisioning system and the service provisioning system for end-user network and service provisioning; the completed service activation activities result in a service update to the billing system where the billing of the end users is activated.

The technical service testing should also include technical service integration with other existing services where appropriate. For example, if the service is bundled with an existing service, then the network and the systems of the two services need to integrate together to form an effective bundled solution.

It is important that the end-to-end technical service testing takes place before operational service testing; otherwise, undue delay will result and the risk of a technical error not being detected at the earliest possible stage will also increase. Some would argue that you could combine the technical service testing activities with the operational service activities. This should not be the case. The operational service testing concentrates on the testing of operational processes and not the technical integration of the service solution. Since the emphasis of the test scenarios and skill sets required for tests execution are different for the two testing activities, it is not wise to mix them together, even though some test scenarios may overlap. However, having the operational personnel to witness the technical service testing will be of benefit for training purposes.

11.1.6 Operational Service Testing

For a successful service launch, it is vital that operational service testing (also known as operational readiness testing) is performed. This ensures everything in the service (network, systems and operational processes) works together as required. This, in turn, reduces operational risks for service operation.

The test scenarios covered in the operational service testing should verify against the system user and operational requirements of the service. An example of operational testing scenarios may involve simulating a fault in the network element (e.g. disconnecting a network connection). This should cause an event in the network element management system and an alarm in the network management system. The operational personnel should see the alarm on the network management system and be able to follow the correct work instruction to diagnose the fault and rectify the fault accordingly.

Before the operational service testing can take place, the following major activities have to be completed:

- end-to-end technical service testing;
- operational processes, including the detailed work instructions, for the service have been defined;
- training on the technology, systems and the operational processes has been completed;
- all the operational/user manuals for the network elements and the systems within the service have been delivered.

Without successful completion of the end-to-end technical service testing, it is difficult to ascertain whether the operational processes developed are complete and fit for purpose. The system may be at error and the operational tests cannot be completed. It is not uncommon to find errors and exceptions on the systems or network elements from operational service testing, as the test scenarios are more process and system-user orientated and, hence, are looking at the service from a different, nontechnical, perspective. One common area where errors or exceptions arise from the service operational testing is the system user interface. The functionalities of the systems may be there, but the usability of these functions may be less than desirable.

Operational processes are very much part of the service solution. Without them, the service will not function. Therefore, one of the entry criteria for the operational service testing stage is clear definition of the operational processes and detailed work instructions. The operational/user manuals for all the network elements and systems are also required. Otherwise, the operation personnel will have no reference material for the network element/system they are managing.

Having the well-defined operational processes and detailed work instructions is only half way to preparing the operational personnel for service launch. Providing them with the training required for using and operating the systems and the network, as well as having the operational processes/work instructions defined, will complete their preparation for service launch. Before running the training activities, it is assumed that the necessary recruitment has been carried out if additional operational personnel are needed to operate the service.

Performing user acceptance testing for all the operational systems should be part of the operational service testing. The end-to-end technical service testing confirms the technical functionality of the service, while the usability of the system is for the system users (normally operational personnel) to decide. The system acceptance tests should be measured against the system user requirements defined (as described in Section 5.2). There is little benefit for the system users if the system functions are not easily accessible for them to perform their tasks. Therefore, system usability requirements and testing should be part of the acceptance criteria of any systems before a service is operational.

Starting the service operational testing without completing any of these activities will increase the risk of service launch failures. If time is the constraint, then one might consider testing different parts of the service in parallel. For example, one can test the order management processes in parallel with that of network management, as both will involve different operational personnel. It is important to validate all the operational areas as part of the operational service testing.

After all the operational processes have been defined and before the start of the service operational testing, it might be a good idea to hold a workshop to have a 'paper walk though' of all the potential service operational scenarios. The representatives of each operational department should have all the respective processes and detailed work instructions ready. For each of the test scenarios there should be operational processes and detailed work instructions defined to deal with the potential operational situation. This 'paper walk through' will reduce the risk of operational service test failures. Any missing processes or detailed work instructions can be identified at an early stage.

At the end of the service operational testing, each operational area will agree (or disagree in some cases) with the readiness for the operation of the service. Launching a service without seeking agreement from all the operational areas involved is disastrous. Without the support from all the operational areas, the service will not be functional. Delaying the service launch will be of a lower risk than launching a service without addressing major operational concerns. Some unhappy customers/end users will result if the major operational concerns are not dealt with before service launch.

11.1.7 Test Exceptions Management

Test exceptions are unexpected events that happen when running through the test scenarios defined in the test specifications; that is, the test results deviate from the expected test results when the tests are run during the service integration activities. Throughout the service integration activities, any test exceptions should be logged and risk assessed before the next stage of integration activities should take place. Any test exceptions or failures result in requirement nonconformances, as the test scenarios/specifications are written based on the requirements defined at the beginning of the development. These test exceptions also signify the noncompliance of service requirements. Some of the test exceptions have to be fixed before further testing activities can take place, and a certain amount of retesting will also be required around the problem areas after the resolutions/fixes have been applied. Assessing the risks of these nonconformances is an important part of the service integration activity.

As part of the service integration strategy, test exception/failure categories should be defined at the start. For example, category 1 failures may be classified as 'service will not function without fixing the exception'. Another way to define the fault categories may involve assessing the potential number of customers/end users being impacted without the test exception being fixed. Depending on the service, an alternative way to define fault categories can be by the amount of potential revenue lost or penalties paid to customers/end users if fault exceptions are not fixed. As a guide, the following definition may be useful:

- category 1 – show stopper;
- category 2 – service affecting, but work around possible;
- category 3 – not service affecting and work around possible;
- category 4 – cosmetic.

Failure categories are used to gauge the success at each stage of testing and should form part of the exit criteria at each testing stage. Completing all the test scenarios does not automatically qualify for proceeding to the next stage of the service integration activities. For example, the completion of service operational testing may not mean that the service is ready for launch. If all the tests within the service operational testing were completed but failed with category 1 failures, then the service is not ready for operation. Remedial actions should be taken for these failures.

After the test exceptions are fixed, they need to be retested to ensure that the expected results are produced and that remedial actions taken for the particular test exception have not jeopardized another part of the solution. Therefore, configuration management of what has been tested and integrated is also important. Configuration management of the test platform is further discussed in Section 11.2.8. For further details on testing and test management, please see *Software Testing and Continuous Quality Improvement* [22].

11.2 Service Integration Strategy

When forming the service integration strategy, you need to define the items that require integrating, the timescale and the resource requirements for all the integration activities. Once the service integration strategy has been defined, the service integration plan can be developed. The service integration plan is dependent on the solution and the timescale in which various parts of the solution are delivered as well as constraints that might be imposed by other services that this service is integrated with. To illustrate how a service integration strategy can be derived, I use the same example service solution as described in Chapter 10 (now featured as Figure 11.3). The implementation strategy for the project does have a strong influence on the service integration strategy, as the project objectives will be common and the service integration strategy needs to fit within the overall implementation strategy.

Within the solution example, the network components to be integrated for the service include the:

- transmission network;
- PS core network;
- access network to customer's main sites via other operators, including the CPE/service-termination device;
- access network to customer's remote sites via other operators, including the CPE at customer's remote sites;
- network connecting network management systems to the network;
- network connecting service management systems to the network;
- network management system network;
- service management system network.

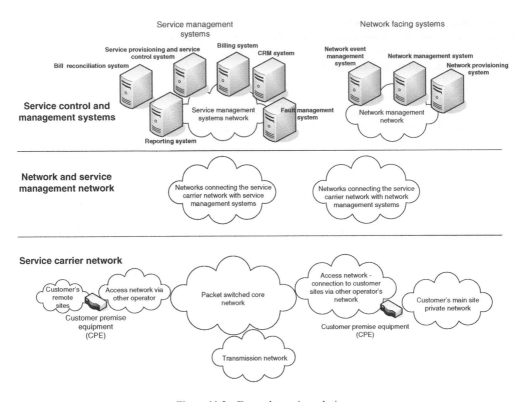

Figure 11.3 Example service solution.

The system solution within this example service involves the following systems:

- new service provisioning and service control system;
- CRM system;
- billing system;
- bill reconciliation system;
- new customer and KPI reporting system;
- fault management system;
- network provisioning system upgrade;
- network element management system;
- network management system.

The operational processes for this service include:

- sales engagement processes;
- customer service – product enquiries processes;
- service and network provisioning processes;
- bill enquiry processes;
- fault management processes;
- network management and maintenance processes;
- service management processes;
- system maintenance processes.

11.2.1 Network Integration

In principle, it is logical to group the various network elements into their respective functions and integrated them accordingly. In this example, the groupings are:

- service carrier network – the network where the service traffic is carried over;
- network and service management networks – the network linking the network layer and the management systems layer;
- network management network – the network where the network element management and network management are connected;
- service management network – the network where the service management systems are connected together.

11.2.1.1 Service Carrier Network Integration

The integration strategy of various groupings also depends on the network design of the various networks. In the example quoted in Figure 11.3, the PS network solution is to work with the access network that is provided by another operator. Hence, interoperability testing will be required between the PS network and the access network. Within that example, the integration of transmission network and the PS network will also be required.

11.2.1.2 Integration of Network Management and Service Management Networks

The integration of the network and service management networks is heavily dependent on the design of the network that the systems are connected to. In principle, the network management systems network needs to communicate with the network elements via the element manager. Therefore, the integration of the network management network with the relevant network elements will be required. This can be independent of the integration of the service management systems and service management systems network.

11.2.1.3 Network Solution Integration

After successful integration of both groupings above, all the network components can be integrated together. End-to-end connectivity from the service management network to the access network, the CPE and the network management network to the CPE and transmission network should be performed to ensure the integrity of network connectivities.

The network integration is summarized and illustrated in Figure 11.4.

11.2.2 System Integration

The logical groupings of the system integration activities can be:

- network-facing systems and
- customer-facing systems (probably more appropriately termed non-network-facing systems);

or

- business support systems and
- operational support system.

In the example described in Figure 11.3, I have decided to group the systems by network-facing systems and customer-facing systems.

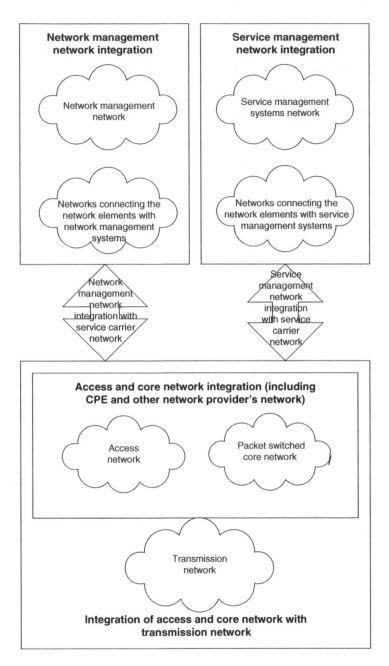

Figure 11.4   Network integration strategy.

11.2.2.1 Network-Facing Systems Integration

Network-facing systems include: network elements/event management system, network management system and network provisioning system. For the network-facing systems, integration of network event/element management systems and network management systems will need to be integrated. This verifies that all the network events are interpreted correctly and that appropriate alarms are displayed for the system users. Integration of a network element with the network provisioning system confirms that appropriate customers' network connectivities can be activated.

11.2.2.2 Customer-Facing Systems Integration

Customer-facing systems include: CRM, fault management, billing, service provisioning and service control, fault management and bill reconciliation system. The integration of the CRM, service provisioning and service control and billing systems should ensure that the service order requirements are fulfilled. The integration of the fault management system, bill reconciliation system, service provisioning system and the billing system with the reporting system will verify that the reporting requirements for the service have been fulfilled.

The system integration activities are illustrated in Figure 11.5.

11.2.3 System and Network Integration

In the example service as described in Figure 11.3, the three major system and network integration activities are between:

- the network-facing systems and all the network components;
- the service provisioning and service control system and the network elements;
- the reporting system and the network elements.

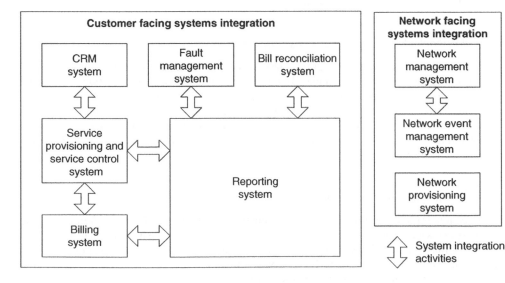

Figure 11.5 System integration activities.

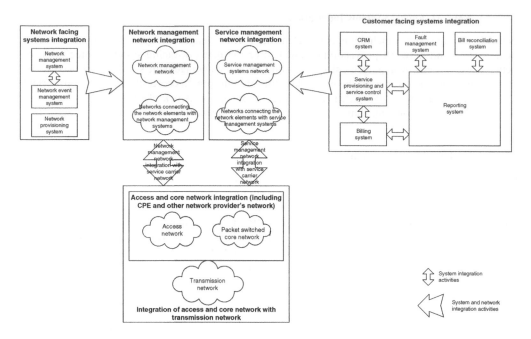

Figure 11.6   Systems and network integration.

These systems are integrated through to the network elements via the systems and network management networks as illustrated in Figure 11.6.

After the initial integration test of the various network-facing systems and network elements, end-to-end technical service testing can take place.

11.2.4 Network, Systems and Process Integration

When all the network and system integration has successfully been performed (i.e. the end-to-end technical service testing has been completed) and when all the appropriate operational personnel have been trained on the operational processes, the integration of the operational processes (i.e. operational service testing) can take place. Various operational scenarios are tested on the operational personnel (e.g. simulation of a fault in the service) to verify the network and system behave as expected and that the operational personnel follow the operational processes and detailed instructions to execute the required tasks.

11.2.5 Service Integration Plan

The service integration plan for this example is illustrated in Figure 11.7. It summaries all of the network, systems and operational integration activities within each of the integration groupings, as well as the different type of integration to take place at each stage of the service integration model described in Section 11.1 and Figure 11.1. Service integration has been divided into two parts: technical service integration and operational service integration. The service is fully integrated only when the operational processes are integrated with the technical environment.

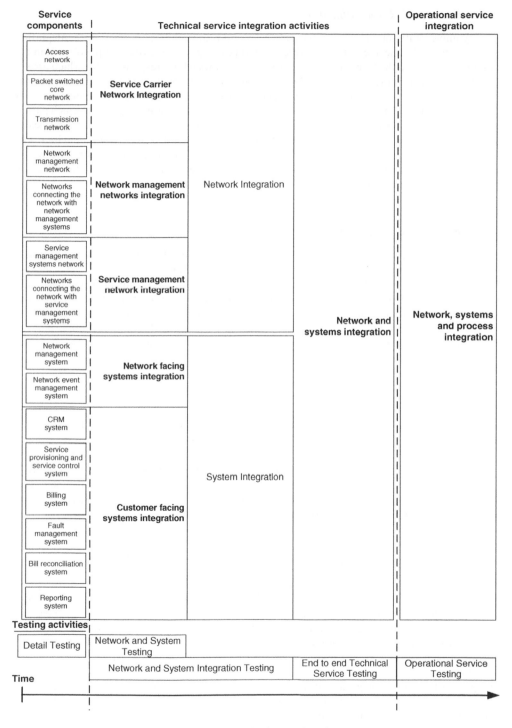

Figure 11.7 Service integration plan.

11.2.6 Roles and Responsibilities

Like any implementation project, defining the roles and responsibilities of the service integration team is essential. All the integration activities should be covered by the relevant people within the team and everyone in the team (and outside the team) knows who is responsible for which area.

As part of the definition of roles and responsibilities, it will be a good idea to draw out the structure of the integration teams. This is particularly useful if the implementation strategy involves third-party suppliers. The team structure should state clearly which team will perform which part of the integration activities and where each party's responsibilities lie. It is essential to define clearly the boundaries of each integration team. Contention may arise for cross-domain activities. For example, who is responsible for the network and system integration piece? The network integration team can easily point their fingers at the systems team and vice versa. The answer to this question depends on the team structure.

11.2.7 Escalation Procedure

The escalation procedure is also one of the elements within the definition of roles and responsibilities. The escalation procedure defines the authorities within the integration team and the escalation paths for events that cannot be resolved between the team members. These escalation paths are normally dependent on the organizational structure of the company. One also wants to have different escalation paths in accordance with failure categories, as category 1 test exceptions may require greater attention than those of lower categories. An escalation procedure is particularly important when unexpected exception scenarios arise where remedial actions involve additional resources, funding or time.

11.2.8 Documentation Structure

Documentation structure should also be part of the service integration strategy. It states what documentation is to be produced within the service integration stage, which area of integration is covered by each document and who will be producing which document. For example, the network integration team will be responsible for producing the detail testing specification of all network elements and test specifications for all the network integration activities. The documentation structure should show all necessary documentation for the service integration activities to be produced with the appropriate authors for each document.

11.2.9 Test Exceptions Management

As described in Section 11.1.7, all test exceptions and incidents are to be logged and exception categories agreed. These test incidents are managed such that the defects are fixed within a reasonable timeframe. Retesting of these failed tests and related tests will require scheduling in the project plan. If certain exceptions cannot be fixed within the timescale required, then this will constitute a nonconformance to one or more requirements. Escalation to the appropriate person may be required. At the end of each test stage, a list of test exceptions is reviewed.

As part of the service integration strategy, it is a good idea to define the tolerance level of test failures before the next stage of service integration activities can take place. For example, one might want to define that no outstanding category 1 (category 1 being the highest risk/highest impact) test exceptions can be carried forward to the next stage of testing.

11.2.10 Entry and Exit Criteria and Output for the Integration Stages

As part of the service integration strategy, it is useful to define the criteria for entering and exiting each integration stage to minimize the risk to the service integration activities. In the event that certain entry criteria are not met but the entry to the next stage of integration is due to start, then risks of not meeting the entry criteria are to be assessed. Plans to mitigate those risks will need to be in place and documented before the next stage should start. One can also decide that the risk is too high to continue and that activities for the next stage are to wait until the entry criteria are met.

In general, entry criteria for each integration stage include (but are not limited to) the following:

- all the exit criteria of the previous stage have been met;
- no outstanding test exceptions from the previous stage (more realistically, probably no outstanding category 1 and 2 test exceptions);
- test environment for the next stage is ready for use and configuration is documented;
- all the test specifications for the next stage are completed and have obtained sign-off from the relevant named authorities as defined by roles and responsibilities within the service integration strategy;
- all the people performing the tests have been trained or have the knowledge to perform the tests.

Exit criteria for each integration stage generally include (but are not limited to) the following:

- all the tests defined for that stage of integration have been completed successfully with no outstanding test exceptions (more realistically, probably no outstanding category 1 and 2 test exceptions);
- all outstanding test exceptions have been risk assessed and resolution dates have been agreed;
- all the requirements nonconformances (as a result of the test exceptions) have been documented and accepted by the appropriate authority;
- test data used for this integration stage has been deleted/archived from the systems under test.

Ensuring the entry and exit criteria are met will minimize the risk in introducing a new service into the stable operational environment.

It is often useful to state the expected input and output at the end of each integration stage. The input for each stage will be similar to that of the entry criteria. Output for each integration stage may include (but is not limited to) the following:

- test results for all the tests performed;
- list of test exception, their respective categories and risk assessment for each exception;
- decision for go/no-go to the next stage of integration.

11.2.11 Test Environment Definition

The test environment of the various phases of service integration is defined as part of the service integration strategy. This leads to activities like seeking of funding and resources in preparation of the service integration phase of the implementation. For example, one may choose to perform all the integration of network and systems in the test laboratory environment before implementing any changes in the live network environment. This is probably wise, as this approach will minimize any potential damage that might be caused by the introduction of new technologies and systems. This also means that a test environment that is representative of the real service will be required, and this can be very expensive.

Within the test environment definition, it is essential to define who is responsible for building and delivering which test facilities. This is especially important if the implementation strategy involves third-party suppliers developing part of the service solution. This is because the suppliers may have a different interpretation of the testing activities and they may also want to avoid any cost that might be incurred by them during the service integration phase. Without a defined test environment for each service integration activity, the integration tasks will not be coherent and more risk will be introduced as a result.

11.2.11.1 Configuration Management

Configuration management of the test environment is very important; especially as each part of the solution may have different software drops at various times for bug fixes and retesting may be required. It also checks if all the integration tests that have been performed are still valid after certain parts of the solution have had a new version of software or configuration installed.

By fixing one problem, the new upgrade may cause failures in other parts of the solution that have been integrated successfully. Without configuration management, it will be difficult to determine where the problem might lie and track what has been tested successfully before the new upgrade.

11.2.12 Regression Testing

After fixing one defect, regression testing needs to take place. This is to confirm that remedial actions have not caused failures in another area within the service solution. For example, during system and network integration, the end-user service was not activated on the network due to an error on the service provisioning system. After fixing this problem, related areas like the network provisioning system, the billing activation of the correct service profile and the network setting for the end user will need to be retested to ensure these functions are still functioning correctly.

11.2.13 Post-Testing Activities

After all the integration activities have been completed and the service platform is ready for service, one needs to check that all the test data is removed and that the configuration for the network and systems has been set for the live service. It is a good idea to rerun some of the end-to-end technical service testing on the live service platform to confirm all the configurations have been set correctly. This is very important, because even misconfiguring one little thing within the solution (e.g. IP address of the back-up network element) can cause major service outages after the service is operational.

11.2.14 Successful Service Integration

To increase the chance of launching the service successfully, it is a good idea to have the operational personnel involved as early as possible; for example, at the end-to-end technical service testing stage or even at the early testing stages or even at the design stage. This way, they will have an early exposure to the technologies and systems to be used. When the systems and the network technologies arrive, there will be less of a surprise. This will also increase the chance of the operational processes working well with the systems and technology being introduced.

In general, it is a better to develop the solution with the involvement of operational personnel as early as possible. This raises awareness of the new service and any operational requirements can be captured at the early stages of the service definition. At the end of the day, it will be the operational personnel who will be running and operating the service, so why not get them on board early? This will also break down the barriers of having to introduce a new service/changes into an already busy operational environment. This involvement does not happen very often and the operation personnel often feel excluded when a new service is launched. For example, getting the operational personnel to take part in the end-to-end technical service testing can form part of their operational training on the service. Of course, resource constraints may prohibit this happening.

11.3 Test Environment versus Live Service Environment

As part of the test strategy for the service, one need to specify the test environment required as well as what type of testing to be performed, as discussed in Section 11.2.10. Since the objective for service integration activity is to minimize the risk resulting from introducing a new service into the stable operational environment, one needs to decide what test facilities to use during which phase of the service integration to achieve this objective.

It is appropriate to perform the testing and integration on a test environment that is representative of the live service environment during the early stages of the service integration (i.e. detail testing, network integration and system integration stages). Beyond that, you need to decide if or when to perform any integration activities in the live service environment. This is heavily dependent on the scale of the service solution and the changes it introduces and, in turn, the risks involved. For a small-scale implementation, I would suggest that the end-to-end technical service testing and operational service testing are to be performed in the live environment where possible.

One of the problems of performing all the service integration activities in the test environment is that, however close the representation of the test environment is to that of the live service environment, there will still be differences. For example, the IP addresses for the network elements and systems in the live environment are likely to different to that of the test environment. Therefore, the configuration for network elements and the systems may need to be changed for the live service. It is small things like these that, if not careful, can prevent the success of the service launch.

It is also a good idea to run the key tests within the end-to-end technical service testing on the live service platform before taking live traffic. This should help to gain confidence that everything on the live service platform has been configured correctly. Any tests that fail at this stage can be a show stopper.

When designing and building the test environment, it will be a good idea to state all the differences between the test environment and the live service environment. With this, the changes to be made are clear when implementing the service in the live environment. The list of differences also highlights the risk areas when switching/implementing the live service.

11.3.1 Integrating with Another Service

Performing the end-to-end service technical testing on the live service platform is especially important if you are integrating the new service with an existing service. Thorough testing of the new service and integration with the existing service in the test environment should take place prior to the integration in the live service environment. The interservice integration ensures both services work harmoniously together in real life. This should also confirm that the existing service is not compromised.

11.3.2 Volume/Stress Testing

Volume testing is the simulation of the service operation under load (i.e. lots of service requests and lots of traffic in the network). This type of testing is essential to ensure the stability of the service platform. The difficulties with this type of testing for a service are having the whole solution (the network and the systems) available together and generating the load to make the tests meaningful. Most suppliers will (should) test the individual network element or system under a specified load condition. However, bringing the whole solution together and putting it under stress may not be as easy. Performing this type of testing before service launch is desirable, if not essential. In addition, having such test platforms is costly. One needs to balance the cost, the benefits and the risks involve for not performing this type of testing. If it has been decided that volume/load testing is not feasible, then ways to mitigate those risks will need to be considered.

11.4 Post-Service Launch Reviews

The other key to the success of launching new services is having the service design team available to support the operational teams for any teething problems during the first few months of service operations. The post-service launch review is a forum where all the operational teams meet with the product manager for the service and the design team to evaluate any operational issues and to review any outstanding test exceptions that are meant to be resolved after the service is operational.

Post-service launch reviews are particularly useful if there are trial customers/end users using the service before higher numbers of customers/end users are taken onto the service. This will give the operational teams a chance to refine their operational processes. This forum also provides the feedback into the design area for future service enhancements for the service.

12

Service Withdrawal, Migration and Termination

Most operators and service providers are only worried about launching new services; when the services come to the end of their useful lives, withdrawing and decommissioning these services is often not their priority. However, effective service withdrawal is essential to avoid unnecessary cost and overheads and will also reduce potential risks and liabilities. This chapter illustrates, with examples, how service withdrawal can be done and the factors to consider during service withdrawal activities. Reference to the eTom process model is also made.

Service migration mainly occurs when the existing service is superseded by the new service, normally due to technology advancement. Customers/end users are migrated to the new service. Service migration can also occur when customers/end users migrate from one service provider to another, as in the case of number portability and broadband end-user migration. In both scenarios, migration activities and steps are to be thought through to minimize risk of failures. In this chapter, the service migration processes and activities are stated and factors to consider are discussed. A service migration example is also given to illustrate with the service migration activities.

Service termination for customers has already been discussed in Chapters 8 and 9. This chapter states all the key events for service termination.

12.1 Service Withdrawal

Service withdrawal is probably the most complicated out of the three scenarios under discussion in this chapter; hence, we will tackle it first. It is important to plan and think through the service withdrawal activities in order that no stone is left unturned and that all relevant parties are consulted. Service withdrawal is essentially reversing everything that was implemented during the service design process. From a conceptual point of view, the order of withdrawal activities goes backwards from the service integration stage back to service components within the service integration plan as illustrated in Figure 11.7. However, each service withdrawal should be considered individually.

Before withdrawing the service, there are six broad areas to assess:

- commercial contracts with customers/end users and suppliers/maintenance and support contract;
- cost of service withdrawal;

- technical details on network and systems components;
- operational support/organization changes;
- interservice dependencies;
- regulatory and finance implications.

Most of the items above should be part of the feasibility study for the service withdrawal. Please refer to Section 12.1.2 for further details.

For those who are familiar with the eTOM [31] model, the process described in this section is the expansion of the 'managed service exit' process (process identifier 1.2.1.5.8).

12.1.1 Service Withdrawal Process

The service withdrawal process is as described in Figure 12.1. This is a combined technical development and business process.

12.1.1.1 Phase 1: Feasibility and Business Approval

The objectives for performing feasibility studies for service withdrawals are to:

- identify items for withdrawal (systems/system elements, network elements, operational processes/departments/groups);
- assess the impacts on other services, network, systems and operational areas as a result of the withdrawal;
- assess any contractual implications for existing customers/end users who are currently using the service;
- consider implications on the support arrangements of the systems and network for the service;
- estimate resource requirements (including third-party involvements);
- estimate cost and timescale of service withdrawal.

Similar to that of designing a service, designing and planning the withdrawal of a service require considerations of all areas involved. Estimated cost of withdrawal should be as accurate as possible, as it is a major input into the business case for service withdrawal. The cost should include the cost for implementation, testing and personnel redundancy (if relevant). A percentage variance on the accuracy should also be stated.

Since the costs and impact will be used in the business case, it is important that the feasibility study has obtained the sign-offs from various impacted departments before the figures are used in the business case.

12.1.1.1.1 Output of Phase 1
Outputs of phase 1 are:

- signed off feasibility study document;
- approved business case;
- budget allocation for service withdrawal.

If the business case is not approved, then the result of the gate 1 review meeting is automatically a 'no go'. No gate review will take place.

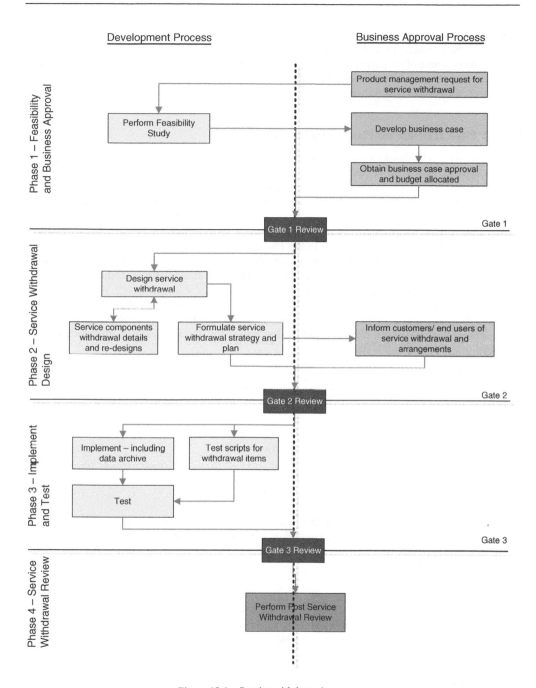

Figure 12.1 Service withdrawal process.

12.1.1.1.2 Gate 1 Review Result

Base upon the completed feasibility study, the business can decide if the service should be withdrawn. The outcome of the gate review meeting can be approval to withdraw service or the service is to remain. The reasons for each outcome should be recorded. After the gate 1 review has approved the project, the budget is allocated for service withdrawal.

12.1.1.2 Phase 2: Design and Plan Service Withdrawal

The feasibility study should provide a good understanding of all items to be withdrawn in the various systems, the network and the operational areas. In the design phase, further details of changes to be implemented are defined. These changes are documented in the technical service withdrawal document. Further details of changes in configuration or design are documented in the respective detailed design/specification document for each system or the network detailed design document. If systems and/or network are to be redesigned as a result of the service withdrawal, then the redesign activities will be performed in this phase.

Customer migration and termination may also take place as a result of the business decision of withdrawing the service. A separate service migration plan and strategy for customers will need to be developed. Please refer to Section 12.2 for details.

A service withdrawal strategy and implementation plan can be developed once the design of the service withdrawal is in a stable condition. This should be available before proceeding to the next phase. Preparation of the test platform and the test scripts for the changes to be implemented can also start once the redesign is stabilized.

12.1.1.2.1 Phase 2 Documentation and Output

Documentation to be produced in the design phase includes:

- technical service withdrawal document;
- detailed design specification documents for all the network elements and systems to be changed;
- service withdrawal strategy and implementation plan;
- network high-level redesign document;
- IT solution/business and operational support system high-level redesign document (if required);
- high-level operational process redesign document (if required).

All the documents should be signed off before entering the gate 2 review. The high-level redesign documents are required if significant changes are introduced. Otherwise, all the redesigns are covered in the technical service withdrawal document. The structure of these documents is as illustrated in Figure 12.2.

12.1.1.2.2 Gate 2 Review Result

Gate 2 ensures that everyone is ready for the service withdrawal activities. This also confirms that all the required resources are available. At the gate 2 review, the result can be approve to proceed to implementation and test phase or to put on hold.

12.1.1.3 Phase 3: Implement and Test

Phase 3 is where the changes for the service withdrawal are implemented and tested. Apart from implementing the changes, test scripts for these changes will also need to be written. These tests ensure that the implementation is successful and that other interrelated services are not affected by the changes made. If changes are significant, then a test platform should be used and the changes

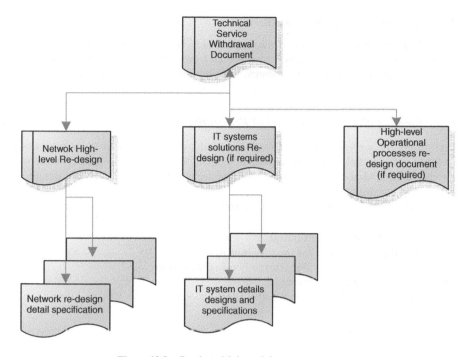

Figure 12.2   Service withdrawal document structure.

tested on the test platform before implementation to reduce operational risks. For details on service withdrawal implementation and testing, please refer to Section 12.1.4.

12.1.1.3.1 Phase 3 Documentation
Documents to be produced in this phase include:

- service withdrawal project plan;
- test platform design (if required);
- test cases and test scripts for all changes made;
- test results;
- list of exceptions.

12.1.1.3.2 Gate 3 Review Result
Gate 3 review confirms that all planned service withdrawal activities have been performed and that no issues or test exceptions are outstanding. This is normally held at the end of the service withdrawal process, unless some major issues arise that will hinder or stop the service being withdrawn. In the latter case, the gate review forum can decide to abandon the service withdrawal and mitigate any associated risks. The review result can either be service withdrawal complete or on hold until major issues are resolved.

12.1.1.4 Phase 4: Post-Service Withdrawal Review

It is suggested that the post-service withdrawal review is to be held 1 month after the service has been withdrawn. This is to monitor any customers'/end-users' activities for the service. All outstanding issues for the service withdrawal should have been resolved by this point.

12.1.2 Performing Service Withdrawal Feasibility

When performing the service withdrawal feasibility study, the first document a service designer should refer to is the technical service solution document, where it should state how the service functions end to end, what service components there are and which other detailed design documentation should be consulted.

Input material into the feasibility study should include:

- business drivers for service withdrawal;
- high-level commercial impact for the withdrawal;
- the number of customers/end users currently using the service;
- current traffic volume of the service;
- any contractual commitments for customer/users and suppliers.

The feasibility study should identify:

- all the technical service components (network and systems) that require withdrawal or redeployment or redesign;
- other suitable services for customers/end users to migrate to (if relevant);
- the operational/organizational impacts or retraining requirements;
- any interservice impacts;
- the customer/end-user impacts;
- the sales and marketing impacts (including commercial contracts to be withdrawn and the marketing material to be withdrawn);
- the network capacity/resource impact;
- any regulatory implications and the impact on the interconnections to other operators;
- any support contract implications;
- any potential risks and issues;
- the assumptions made when performing the study;
- a rough timetable for the withdrawal, with estimated lead times for each of the withdrawal items;
- all the costs and resources required for the withdrawal.

As part of the feasibility study, representatives in the supplier management area should be involved to review the existing commercial contracts for:

- customers'/end-users' commitments;
- network element maintenance;
- systems maintenance;
- other network operators that may form part of the service.

All the above should be reviewed for contract conformances. Resources from the network and systems suppliers may be required for the service withdrawal activities. Additional supplier negotiation may be necessary for equipment recovery.

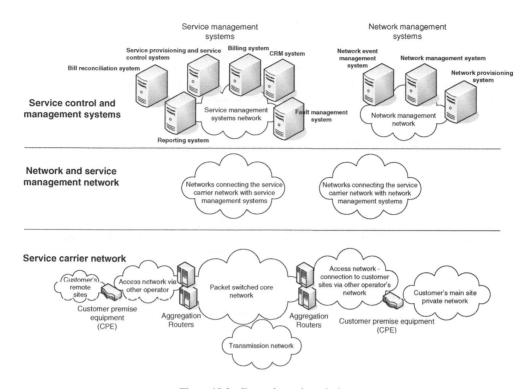

Figure 12.3   Example service solution.

12.1.3 Design and Plan Service Withdrawal: An Example

To fully explain the service withdrawal activities, I have used the example service solution in Chapter 11 to illustrate the potential items for withdrawal. I have included more detail in the service solution example for this service withdrawal scenario. As before, generic terms of the network and systems are used rather than naming specific network devices and systems in order that the readers can easily adapt the principles into their services. The example solution is as Figure 12.3.

In this example solution, the service demarcation points are the CPE at a customer's remote sites and the CPE (service termination device) at the customer's main site. The service to be withdrawn is one of the managed IP VPN data services within the service solution that was integrated in Chapter 11. The service being withdrawn is designed specifically for small/medium-sized business customers using low-end CPE with a low-speed (typically 2 Mbps) leased-line access network provided by the alternative network provider (OLO). Since the cost of the access network has reduced significantly as a result of DSL technology, this service has been superseded by the broadband/ADSL service.

12.1.3.1 Commercial Contract Components

The commercial contracts that need attention include:

- Review of customer contract. We need to ensure contractual commitments have been fulfilled.
- Contracts with the alternative network provider for the access network. This is ordered on a per customer site install basis. Therefore, we need to confirm if the minimum terms of these contracts have been met. Otherwise a penalty will be liable.

- Network element contract. Since the CPE is no longer useful for other services within the service portfolio, negotiation with the supplier for a buy-back arrangement should be attempted.
- System maintenance contract. The reporting system needs to be decommissioned. We need to review the maintenance contract to confirm if that is included as part of the service package. Otherwise additional cost will be incurred. One also needs to confirm if the maintenance contracts for the rest of the systems support the withdrawal of service.

12.1.3.2 Customer-Facing Components

Customer-facing activities that need to take place before service withdrawal include:

- Informing the customers of the service withdrawal.
- Drafting a customer service migration plan, if migration of customers to other/new services is required. Customer migration will need to take place before the service can be withdrawn.
- Service termination activities if customers decide to terminate their service. Customer service terminations are to take place before the service can be withdrawn.
- All the CPE from the customer sites is recovered and management connections to the network element management and network management system are disconnected.
- Marketing and sales literature for the service will need to be removed.
- Removal of service content and links on the marketing websites.

12.1.3.3 Network Components

The network for service can be broken down into the following components:

- transmission network;
- PS core network, with aggregation routers used exclusively for this service;
- access network to customer's main sites via other operators, including the CPE/service termination device;
- access network to customer's remote sites via other operators, including the CPE at customer's remote sites;
- network connecting network management systems;
- network connecting service management systems;
- network management system network;
- service management system network.

12.1.3.3.1 Network Impact

The network components that require actions include:

- The support of the CPE will be withdrawn. This CPE is no longer in use for other services. All CPE should have been recovered as part of the customer migration/termination activities.
- The aggregation routers are decommissioned and disconnected from the PS core network. These routers are redeployed in other parts of the network for another service.
- The network connections from other network operators to the aggregation routers need to be disconnected and the capacity released.
- The transmission network connecting the aggregation routers to the core routers needs to be disconnected and the network capacity released. Transmission redesign work is required.

- The network connections between the network element management system and the aggregation routers are to be disconnected. The impact on the design of the network management network should be minimal.
- The network connecting service management systems to the aggregation routers needs disconnecting. The impact on the design of the service management network should be minimal.
- The network management connection between the CPE and the network management systems should already be disconnected as part of the customer migration activities.
- The service management connection between the CPE and the service management systems should already be disconnected as part of the customer migration activities.

Since other VPN services are sharing the same PS core network equipment and the transmission network is connected to the core network, those should remain in place. The CPE should be withdrawn as part of the customer migration or customer termination activities. As the CPE is taken out of service, the management path from the CPE to the network management system should also be disabled. Please see Sections 12.2 and 12.3 for details on customer migration and termination respectively.

12.1.3.4 System Components

The system solution within this example service involves the following systems:

- service provisioning and service control system;
- CRM system;
- billing system;
- bill reconciliation system;
- fault management system;
- network provisioning system;
- network element management system;
- network management system;
- reporting system.

The service has been configured into all the above systems and the systems are shared with other services, apart from the reporting system. The reporting system is exclusively for this service. It takes information from the fault management, service provisioning and service control system and the customer CPE to generate reports for fault management, service management and network performance respectively. It also provides a secure Web interface for the customers to download the reports and view their monthly bills. There is no external system interface to other network providers for the service. All access network orders are done manually using faxes.

12.1.3.4.1 System Impact
Actions to be taken with the systems are as follows.

- Service provisioning and service control system.
  - The service, the service profile and the control attributes for this service are to be disabled.
  - One needs to ensure that other services sharing the systems are not affected after the service disablement (e.g. the interface to the network elements).

- CRM system.

  - ○ The service is to be disabled in the CRM system.
  - ○ Ensure that no new customer can order the service.
  - ○ Historical customer data of the service should be archived.

- Billing system.

  - ○ The service should be disabled from the system and the price plan withdrawn.
  - ○ Check if all the customers on the service have been terminated or migrated to other services.
  - ○ Confirm no new customers can be enabled for this service. Care must be taken to ensure that the existing customers that are using other variants of the IP VPN service are not affected by the change.
  - ○ It may be wise to review outstanding bills for this service.

- Bill reconciliation system.

  - The service should be removed from the system.
  - It is worth checking if there are any access network charges outstanding for this service, as all the customer access network connection should have been terminated.

- Fault management system.

  - ○ The service should be removed from the fault management system.
  - ○ There should be no outstanding faults on the service. Any faults should be noted and the impact assessed. If the fault is on the network element to be decommissioned, then diagnosis of the problem should be performed be before decommissioning. The diagnostic is necessary because the network element is going to be redeployed.

- Network provisioning system.

  - ○ The CPE is only used for this service. The CPE and the provisioning script for the CPE should be removed from the network provisioning system.

- Network element management system.

  - ○ The MIB files that are only related to that CPE to be withdrawn should be removed.
  - ○ The aggregation routers should be removed from the system as part of the decommissioning of the service.
  - ○ Confirm if the CPE has been removed from the system as part of the customer migration/termination.

- Network management system:

  - ○ The aggregation routers and the CPE should be removed from the network map/network inventory within the network management system.
  - ○ If the service was configured in the network management system, then the service is to be removed.
  - ○ It is worth checking that the current network map/network inventory is not affected by the removed of these network elements.

- Reporting system.

  - ○ The Web interfaces and the links to the company's website relating to this service are to be removed.
  - ○ Security to the company's other Web services is to be assessed to ensure that the security access granted to the customers to this service will not enable them to access other services that are not appropriate to them.

o Data and the reports in the system should be archived.
o Data feeds from the network elements, billing system, fault management system and the service provisioning and service control systems for this service should be disabled. This disablement should not affect the data originating systems/network elements or their data stores.
o The system is to be disconnected from the service management network.
o The system is to be decommissioned. This includes the removal of the system from the service management system network.
o System hardware is to be reused for other systems after some hardware upgrades.

12.1.3.5 Operational Components

The following operational groups are looking after this service:

- sales engagement;
- customer service–product enquiries;
- service and network provisioning;
- bill enquiry;
- fault management;
- network maintenance;
- service management;
- system maintenance.

The operational teams for customer service and bill enquiry are dedicated to this service. Other operational groups are shared with the rest of the IP VPN service portfolio.

12.1.3.5.1 Operational Impact
All the operational processes for this service (within all the operational areas listed above) will need to be withdrawn and all the detailed work instructions removed. The customer services and bill enquiry teams need to be retrained to look after other services. Some head counts are lost; hence, some of the team members will be made redundant or redeployed to other operational areas. Since the rest of the resources are shared with other services, they are 'absorbed' within their respective operational groups.

12.1.4 Implementation and Testing of Service Withdrawal

12.1.4.1 Implementation

The implementation of service withdrawal activities should be done when the network/system is in its quietest period. It is good practice to back up the existing configuration, software and data before changes are made. In the event that the changes have adverse effects, the original operation can be restored. It is also wise to make one change at a time within one system/network area. This will help with the isolation of faults/failures as a result of the service withdrawal activities. Configuration management of all the systems and network elements is an important item to manage as part of the service withdrawal implementation.

12.1.4.2 Testing and Test Platform

For each technical service withdrawal activity, testing has to be performed on the reference network or systems that are representative of the real operational environment before implementation. This

ensures that the changes (the withdrawal) introduced do not affect other services or other functionalities of the systems and network elements associated with the withdrawal. This will minimize
the risk of failures. This testing is especially important for the new/changed network configurations
introduced as the result of the service withdrawal.

In the example above (Section 12.1.3), the following areas for testing should be included:

- connectivity and routing/resiliency testing in the PS core network after the removal of the
  aggregation routers;
- testing of the network management and network element management systems as a result of
  the removal of aggregation router and the CPE;
- security testing as a result of the removal of the external interfaces to the reporting system;
- functionality and network provisioning testing of other services on the network provisioning
  system after the service has been removed;
- functionality and service profile testing of other services on the service provisioning and service
  control system after the service has been removed;
- functionality and configuration testing of the billing system ensuring that disabling of the
  service and price plan do not affect other related services;
- functionality and configuration testing of the bill reconciliation system to confirm that
  removal of the service does not affect other services on the system;
- removal of service from the fault management system – ensuring no faults can be logged
  against the service and that other related services are not affected by the change;
- interfaces to the reporting system – ensuring all the interfaces to the reporting system have
  been closed and that no further data (from all the systems) is generated for the service.

12.1.4.3 Test Exceptions

Test exceptions are unexpected events that happen when running through the test scenarios defined in
the test specifications; that is, the test results deviate from the expected test results during the service
withdrawal activities. All test exceptions should be logged and risk assessed. Since the change is being
implemented in an operational environment, these test exceptions may have a detrimental effect on
other services. Therefore, assessing the risks before proceeding with the change or making the
decision not to implement the change or reverse the change implemented is very important.

Test failure categories for service withdrawal activities can be the same as those of service
integration. As a guide, the following definition can be used:

- category 1 – show stopper;
- category 2 – service affecting, but work around possible;
- category 3 – not service affecting and work around possible;
- category 4 – cosmetic.

Since the change is implemented in the operational environment, it is recommended that test
failures with categories 1 and 2 are of high risk and these changes should not be implemented
until the problems have been resolved and retested.

12.1.5 Service Withdrawal Strategy and Service Withdrawal Plan

As for all strategies, one needs to define the objectives. In general, the main objectives for the service
withdrawal strategy are to minimize operational risks and to have minimal disruptions for the
customers/end users. With this in mind, the service withdrawal activities should start from the

customer-facing activities and commercial contract items, followed by the removal of external interfaces. Once all the external factors have been taken care of, the decommissioning and removal of the network and system components can take place. Operational areas will be dealt with last, as you need the operational personnel to help with the service withdrawal activities. However, this does not mean that the retraining of operational personnel to support other services cannot take place in parallel.

To illustrate the above points, the example solution in Section 12.1.3 is used. This example also includes the service withdrawal activities for the service withdrawal plan. It is assumed that all the customers have either been migrated to other services or have been terminated from the service. For customer migration and customer termination, please see Sections 12.2 and 12.3 respectively for details.

12.1.5.1 Decommissioning of External Interfaces

As stated earlier, all the activities for the customer-facing components (as stated in Section 12.1.3.2) and the commercial contract components (as stated in Section 12.1.3.1) should be performed before the external interfaces are withdrawn. Therefore, all the commercial, customer and supplier contractual commitments are reviewed and their impact assessed before starting any service withdrawal activities.

The first area of service withdrawal is the decommissioning of external interfaces. This is to minimize the exposure of external influences. In the example service (as described in Figure 11.3), activities to be performed to close down all the external interfaces include:

- removal of all marketing and sales literature of the service;
- removal of service content and links on the marketing website;
- inform customers of the service withdrawal;
- withdrawal of all access network interfaces to other network operators that are related to this service;
- removal of the Web interfaces and links to the reporting system;
- inform all customer-facing operational areas that the service is to be withdrawn and customer enquiries are to be dealt with appropriately.

12.1.5.2 System and Network De-Integration

Once all the external interfaces are removed, the next step will be the de-integration of the system and network interfaces. By decoupling the network and system components, you are removing the interdependencies between the two domains. This enables the network and the systems to be decommissioned independently of each other. This will improve the efficiency of the service withdrawal plan, as the network and system service withdrawal activities can be carried out in parallel.

The three major system and network de-integration activities are between the:

- network provisioning system and the network elements;
- network-facing system and all the network elements;
- service management system and the network elements.

In the example service described in Figure 11.3, the components that require de-integration include:

- The disconnection of the network connection between the network element management system and the aggregation routers.

- The disconnection of the network connecting service management systems (i.e. service provisioning and service control system) to the aggregation routers.
- The decommissioning of the data feeds from the network elements to the reporting system.
- The disconnection of the reporting system from the systems management network.
- Checking that the network management connections between the CPE and the network management system have been disconnected as part of the customer migration/termination activities.
- Checking that the service management connection between the CPE and the service management systems (i.e. the service provisioning and service control system and reporting system) are disconnected. This should already be disconnected as part of the customer migration activities.

These de-integration activities are illustrated in red in Figure 12.4.

After the de-integration of the systems and network components, the various network and system decommissioning activities can take place.

12.1.5.3 Network Decommissioning

In general, the logical groupings of network elements for decommissioning are to disconnect:

- the service carrier network from the network management network;
- the service carrier network from the service management network;
- the network management network from the network connecting the network management systems;
- the service management network from the network connecting the service management systems.

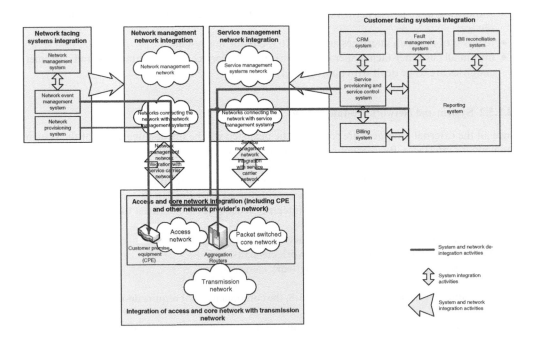

Figure 12.4   Systems and network de-integration.

In the current example, the CPE should have been disconnected from the service and network management network as part of the customer service termination or migration. The other network component that requires decommissioning is the aggregation router. Therefore, the decommissioning activities for the aggregation routers include:

- disconnecting the aggregation routers from the PS core network and the transmission network.
- disconnecting the aggregation routers from network management and service management network (if not already done so as part of the actions in Section 12.1.4.2).
- decommissioning and powering down of the aggregation routers.
- removing related network elements in the network element management, network management system and network map/inventory.
- reconfiguring the PS core network and the transmission network as per the redesigned network document.

In this example, other services are sharing the network management and service management network and systems. Hence, these and other network connections should not be touched.

12.1.5.4 System Service Withdrawal

From a strategy point of view, the logical sequence of withdrawing the service from the systems is:

- de-integration of interlinks between the systems;
- then withdraw service from the customer-facing systems;
- followed by withdrawing service from the network-facing systems.

In most cases, once the interlinks between the systems have been removed, the service can be withdrawn from all the systems at the same time, provided that no active customers are on the service and that resources are available to do so.

12.1.5.4.1 De-Integration of Interlinked System

In the example given, the only system requiring de-integrating is the reporting system. Therefore, the interfaces between the reporting system and the fault management system, bill reconciliation system, service provisioning system and the billing system require de-integrating, as illustrated in red in Figure 12.5. The rest of the systems are shared with other services; hence, the physical de-integration of the system interlinks is not possible and not necessary and should remain integrated.

12.1.5.4.2 Withdraw Service from Customer-Facing and Network-Facing Systems

The service should be disabled from the following systems:

- CRM system – the service is to be disabled in the system;
- service provisioning and service control system – the service profile and the control attributes for this service are to be disabled;
- billing system – the service should be disabled from the system and the price plan withdrawn;
- bill reconciliation system – the service should be removed/disabled from the system;
- fault management system – the service should be removed from the fault management system;
- network provisioning system – the service should be disabled from the system;
- network element management system – only the MIB that is related to that CPE to be withdrawn should be removed.

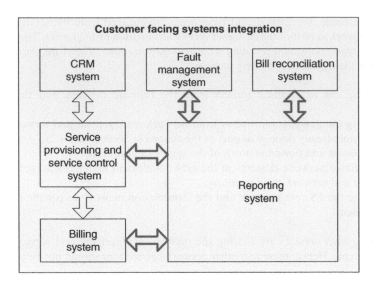

Figure 12.5   System integration activities.

Care must be taken not to delete the service such that previous records of customers and data on the service cannot be retrieved when necessary.

12.1.5.4.3 Decommissioning of Reporting System

The remaining decommissioning activities for the reporting system include:

- archiving data and reports;
- disconnecting the reporting system from the service management network (if not done as part of Section 12.1.4.2);
- decommissioning and power down the system;
- removal of hardware for upgrading.

12.1.5.5 Operational Service Withdrawal

All operational processes and detailed work instructions associated with this service should be marked obsolete. Any operational processes for other services that are dependent on this service will need to be revised. In this example, no other service is dependent on the service being withdrawn. Therefore, all operational processes and detailed instructions associated with this service are marked obsolete.

Retraining of personnel in the operational area should start when business approval has been obtained for service withdrawal. Those operational personnel that are to be made redundant should be told at the start of the service withdrawal process.

12.1.5.6 Service Withdrawal Plan

The service integration plan for this example is as Figure 12.6. This summarizes all of the service withdrawal activities and the sequence of events for service withdrawal.

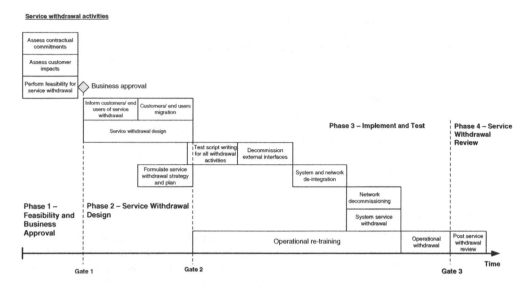

Figure 12.6   Service withdrawal plan.

From the project plan point of view, most of the activities start after business case approval, when the budget for the withdrawal activities is allocated. Once the business has approved the withdrawal of the service, the customers/end users should be informed. A customer/end-user migration plan should be drawn up detailing when customers/end users are migrated to other services (please see Section 12.2 for details on migration).

In parallel with the service migration activities, service withdrawal activities are designed. When the high-level activities are formulated, the service withdrawal strategy and plans are drawn up. Towards the end of the service withdrawal design stage, when details of service withdrawal tasks are known, one may consider designing the test platform and writing the test scripts and specification to ensure that the service withdrawal tasks have been completed successfully in advance of gate 2 approval. This may help to save some time in the project plan.

When all the test scripts for the service withdrawal activities have been signed off and gate 2 approval has been gained, decommissioning of various systems and network components will start. As per the strategy mentioned above, the external interfaces are to be decommissioned first, follow by the de-integration of the system and network elements. Network decommissioning and system service withdrawal activities should happen after the separation of system and network elements.

Operational retraining for other services can take place anytime after gate 2 approval. Operational withdrawal of the service should take place after all the technical components of the service have been successfully withdrawn. Lastly, post-service withdrawal reviews should take place a month after the service withdrawal activities have been completed.

12.1.5.7 Roles and Responsibilities

Defining the roles and responsibilities of the service withdrawal activities is especially important, as the project is probably not the most exciting thing to do. Seeking an individual's support may be difficult, as some of the operational personnel may be made redundant as a result of the service withdrawal. Named resources from each operational area as part of the team structure are essential; otherwise service withdrawal activities will not happen. The team structure should state clearly which team will perform which part of the service withdrawal activities. It is crucial to define clearly the boundaries of each team, as contention may arise for all the cross-domain activities.

12.1.5.7.1 Escalation Procedure

The escalation procedure is also an important element within the definition of roles and responsibilities. The escalation procedure defines the authorities within the service withdrawal team and the escalation paths for issues that cannot be resolved within the team members or for inactions of certain activities. The escalation procedure should also include the escalation of test failures.

12.1.5.8 Exceptions Management

As described in Section 12.1.4, all test exceptions and incidents have to be logged and exception categories agreed. These test incidents need to be managed such that these failures are fixed within a reasonable timeframe. Retesting of these failed tests and related tests will be required. If certain exceptions cannot be fixed within the timescale required, then escalation to the appropriate person may be needed.

12.1.5.9 Entry and Exit Criteria and Output for Each Service Withdrawal Stage/Activity

As part of the service withdrawal strategy, it is useful to define the criteria for entering and exiting each service withdrawal stage to minimize the risks. Typically, the entry criteria for each stage of the service withdrawal process will include: the design for all the activities within that stage has been signed off; all the resources are trained and available, the network elements or the systems are available and so on. The general exit criteria for each service withdrawal stage will include: the all the activities within that stage have been completed successfully and any exceptions are logged and risks assessed.

For service withdrawal, it might be more appropriate to further define entry and entry criteria for each service withdrawal activity within the implement and testing stage. This will ensure the risks are minimized for each service withdrawal activity.

12.1.6 Post-Service Withdrawal Review

The post-service withdrawal review should take place 1 month after all the service withdrawal activities are completed. This is to ensure that no outstanding exceptions and issues remain.

12.2 Service Migration

Service migration normally occurs when customers/end users want to move from one service to another. In some instances, customers/end users want to move from one service provider to another; for example, changing their broadband service provider. Interservice provider migration is a very complicated subject. It is dependent on the technology, the service and many regulatory issues. I have no intention of covering it in this book.

Another service migration scenario could be the result of selling the business unit for the service to another provider or the result of merging two service providers. In both scenarios, the customers/end users will need to be migrated from one service platform to another.

Some people might say, 'Why bother with service migration? It is much simpler to "cease and reprovide" the service'. The main problems with terminating the customers'/end-users' service and reproviding it are service outages and the potential issue with minimum term of service in the contract. Above all, having service outages and contractual issues are hardly good service experiences for the customers/end users, especially if these are not caused by customer/end-user actions.

The principle for service migration is to minimize customer/end-user impact. With this in mind, one needs to ask:

- Is there any pre-migration work that can be carried out for the customers/end users before the migration activities? For example, can the customers/end users be enabled in all the new service components before migration, with the customers'/end-users' service migration, therefore, becoming just a 'flick of a switch'?
- What is the acceptable outage window for the migration?
- Are all the customers/end users being migrated or just a proportion of them? What are the criteria for migration?
- Will having both services (the one being migrated from and the one being migrated to) running in parallel for the customers/end users be possible/necessary? Will that gain any benefit in terms of the customer/end-user experience? What are the risks of doing so?
- Should you migrate all the customers/end users at the same time or should you split the migration into reasonably sized batches? If they are splitting it into batches, what are the criteria for the split? Geographical or just random?
- Is there enough network and system capacity to accommodate the migrated customers/end users?
- Do the customers/end users need different CPE for the new service?
- Do the customers/end users have a choice of which service to migrate to?
- Are there sufficient operational resources to cater for the migrated customers/end users?
- Do you need to build additional/temporary systems resources/support for the migration?
- How do you know if the migration has been successful? What tests do you do to prove that the migration has been successful? Do you just wait for customers/end users to log faults if the migration does not work?
- What are the success criteria for the migration? Are there any KPIs for service migration?
- How about billing? Are there price differences between the two services?
- How about contractual issues? Will the customers/end users need to sign a new contract for the new service? Will the terms and conditions be the same?
- After the migration, does this mean that the minimum term for the customers/end users starts again? Are the customers/end users aware of that?

12.2.1 Service Migration Process

The business process for service migration is very similar to that of the service withdrawal process, as illustrated in Figure 12.7.

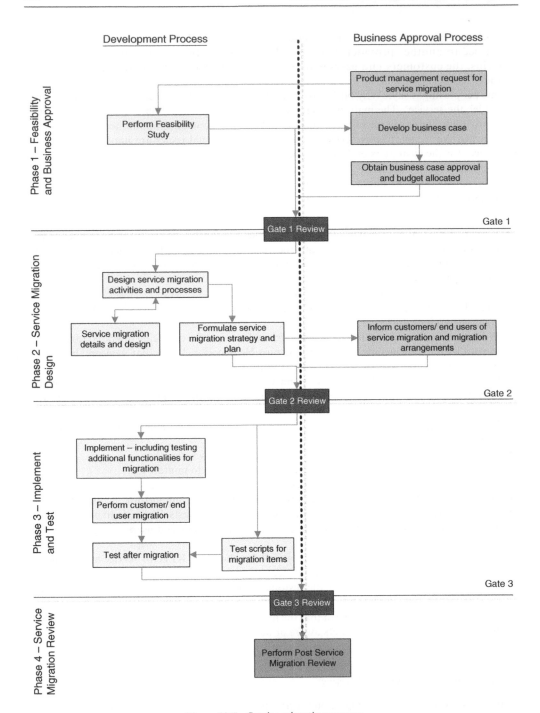

Figure 12.7   Service migration process.

12.2.1.1 Phase 1: Feasibility and Business Approval

The objectives of the feasibility study for service migration are to:

- identify customers/end users for migration;
- identify a service platform for migration;
- investigate the technical differences between the existing service platform and the new/ migrated service platform;
- evaluate the tasks to be performed for each customer/end user;
- identify if there are any pre-build service components for all the customers/end users (e.g. preprovision the network capacity, service profiles) to minimize risk and time during the migration;
- review customer/end-user contractual terms and potential contractual issues;
- assess the implications on network and systems capacity and functionalities as a result of the migration;
- assess the impact on operational areas as a result of the migration;
- identify the time window for migration (when customer/end-user traffic is low);
- estimate resource requirements;
- estimate cost and timescale of service migration activities;
- recommend on the approach to service migration.

Since the estimated costs will be used in the business case, it is important that the feasibility study has obtained sign-offs from the appropriate authorities before the figures are used in the business case.

12.2.1.1.1 Input for Phase 1
Input for phase 1 includes:

- business drivers for service migration;
- technical service solution document for the service that the customers/end users are currently using;
- technical service solution document for the service that the customers/end users are migrating to.

12.2.1.1.2 Output of Phase 1
Outputs of phase 1 are:

- signed-off feasibility study document;
- approved business case and budget allocation for service migration.

If the initial business case is not approved, then the result of the gate 1 review meeting is automatically a 'no go'. No gate review will take place.

12.2.1.1.3 Gate 1 Review Result
Based on the completed feasibility study, the business can decide if service migration should go ahead. The outcome of the phase review meeting can be to approve service migration or for the customers/end users to remain on the current service. The reasons for each outcome should be recorded. After the gate 1 review has approved the project, the budget and resources are allocated for the service withdrawal activities.

12.2.1.2 Phase 2: Design and Plan Service Migration

The feasibility should have identified all the tasks to be performed as part of the service migration. In the design phase, further details of migration tasks are defined and migration plans and strategy are formed. Below are the activities to take place during the design phase:

- design additional network and system capacity (if required);
- design additional functionalities in the systems to support the migration activities;
- define the tasks to be performed for each customer/end-user migration;
- produce customer network redesigns as a result of service migration (if required);
- define the sequence of tasks to be performed and processes required for each customer/end-user migration;
- define the operational processes for migration (if different from that of the service);
- define tests to ensure successful customer/end-user migration;
- form migration plan for the migration phases (if any);
- inform the customers/end users regarding the service migration activities and the potential service outage dates and times;
- test all the configuration/design for migration;
- train operational personnel on the migration processes and work instructions;
- conduct service operational testing on the technical migration and service migration processes and work instructions.

12.2.1.2.1 Phase 2 Documentation and Output

Documentation to be produced in the design phase includes:

- detailed design specification documents for all additional functionalities in the system to support the migration;
- detailed design documents for additional network and system capacity (if required);
- customer network designs (if required);
- operational processes and detailed work instructions for all the service migration tasks;
- test specification for additional network capacity to support the migration (if required);
- test specification for additional functionalities in the systems to support the migration activities;
- test specification for service operational testing on the technical migration and service migration processes and work instructions;
- test specifications for tests to be performed after customers/end users have been migrated;
- service migration strategy and implementation plan.

All the documents should be signed off before entering the gate 2 review.

12.2.1.2.2 Gate 2 Review Result

Gate 2 ensures all the design and test specifications are signed off and that everyone is ready for the service migration activities and that all appropriate operational personnel have been trained.

The gate 2 review also ensures that all the required resources are available. At the gate 2 review, the result can be approval to proceed to implementation and test phase or to put on hold.

12.2.1.3 Phase 3: Implement and Test

The implementation phase is split into stages: implementation of the required infrastructure to support the migration and the implementation of the customer/end-user migration. Below are the activities to be implemented before the customer/end-user migration takes place:

- build and test additional network and system capacity (if required);
- build and test additional functionalities in the systems to support the migration activities;
- perform service operational testing for the technical migration and service migration processes and work instructions;
- prebuild all service elements for all customers/end users where possible (e.g. preprovision the network capacity, service profiles).

Any test exceptions and outstanding issues from the testing activities for the network, systems and operational processes should be reviewed. To proceed to the next stage of implementation (i.e. performing the customer/end-user migration), no outstanding category 1 and 2 test exceptions should be present.

The second stage of the implementation phase is where the migration activities happen. After each customer/end-user migration, a set of tests should be performed to ensure that the migration is successful. This will minimize the potential faults being logged by customer/end users as a result of the migration activities.

12.2.1.3.1 Phase 3 Documentation
Documents to be produce in this phase include:

- service migration project plan;
- test result for additional functionalities in the systems to support the migration activities;
- test result for additional network capacity to support the migration (if required);
- test result for service operational testing on the technical migration and service migration processes and work instructions;
- test cases and test scripts for customer/end-user service migration activities;
- test results for customer/end-user service migration activities;
- list of exceptions for all the tests listed above.

12.2.1.3.2 Gate 3 Review Result
The gate 3 review ensures that all planned service migration activities have been completed successfully and that no issues or test exceptions are outstanding. If the migration activities are divided into different phases, then reviews should be held at the end of each phase to ensure that any improvements on efficiency can be realized and that all the processes and the defined steps for migration are working as planned. Progress reviews should be held throughout the migration period. This is to confirm that the migration is on track and that no major issues arise that will jeopardize the migration plan.

The gate 3 review result can either be service migration complete or on hold until major issues are resolved.

12.2.1.4 Phase 4: Post-Service Migration Review

It is suggested that the post-service migration review is to be held 1 month after all the customers/ end users have been migrated. This is to monitor whether there are issues arising as a result of the migration and if there are any outstanding issues from the migration to be resolved. In addition, any additional system functionalities that are built specifically for the migration should be withdrawn by this point. All outstanding issues from the service migration activities should have been resolved.

12.2.2 Service Migration Strategy

The service migration strategy defines the approach to the service migration and how the migration activities are going to take place. It identifies the interdependencies between the migration activities and the roles and responsibilities of the migration team. The entry and exit criteria for each migration stage, the escalation procedure for issues, the exception management of the test results, the review points and the review schedule should also be stated in the migration strategy.

The migration strategy for the end users is very different from that of the business customers, as business customer migration will involve migrating different sites at different times and will require coordination and project management for each customer. You also need to decide whether these customers' migrations can take place in parallel and assess the risks involved in doing so.

For end-user migration, the migration strategy can be done by geographic area, or in alphabetical order in sensibly sized batches, depending on the service involved. It is wise to have a few days between end-user migration batches. This will give the migration team a chance to fix any recurring faults that result from the migration and modify any scripts/configuration/design if required.

It is also a good idea to start the migration with a small batch of customers/end users to trial the tasks and process defined. This ensures that all the defined tasks and processes work on a small scale and that lessons can be learnt before larger numbers of customers/end users are migrated. This will minimize the risk of failure for the subsequent migration phases/batches.

Monitoring the fault rate for the migrated customers/end users during the migration phase provides a good indication for the success/failure of the migration activities. If there is a common fault from these customer/end users, then the migration design/steps will need to be reviewed/modified to ensure these fault occurrences can be minimized.

12.2.3 Service Migration Example

Using the same example as service withdrawal in Section 12.1 (Figure 12.3), all the customers have either been migrated to the broadband service or been terminated from the service before service withdrawal can start.

12.2.3.1 Service Migration Feasibility Study Example

It has been identified that there are 60 customers (with a total of 285 sites) that want to migrate to the broadband VPN service and the rest of the customers are terminating the service. The technical differences between the two services are the access network, the line speeds and the service profiles for each customer. The following migration tasks for each migrating customer have been identified:

- Perform network redesign for each customer. Some network redesign work will be required for each customer as the service profile and the core network configuration for the customers are going to be different.
- CPE replacement. New CPE is required for each customer site as the existing CPE does not have the correct network interfaces for broadband lines. The old CPE will need to be recovered from the customer sites. The configuration of the new CPE can be preloaded and tested before

the customer engineering visits. This will minimize installation failures on site, hence saving time during the migration.

- Change of service profile for customers. Owing to the change of the service, all the customers adopt the service profile for the broadband VPN service that is already in place.
- Core network reconfiguration. Depending on the redesign of the customer's network, it is expected that some reconfiguration for the customer's core network will be required.
- Order broadband access network connections from alternative network provider for each migrated site.
- Order new CPE for each customer.
- Change service on CRM. Once migrated, the customer profile in the CRM system should reflect the change. The history of the customer should show which service they have previously had and what service they have now.
- Data-fill reporting system for the new service. This should be part of the provisioning process for the broadband service.
- Billing. Once migration has been successful, billing for the broadband service should be activated.
- Reuse old CPE. The CPE recovered from the customer sites will need to be upgraded and can be reused for other VPN services.

In this example, it is possible to prebuild the access and core network configuration before the migration, as a different access network is used and a different set of aggregation routers is going to be used. No additional service migration test is required. The existing service provisioning tests for the broadband service is sufficient.

There is sufficient network and system capacity to cope with the migrated customers. Hence, no additional system support functionalities are required.

All the customers need to sign a new contract for the migrated service, as there is a minimum term for the broadband access network connection imposed by the alternative network provider and the pricing for the new service is different.

Most of the provisioning processes for the service to be migrated to are treated as new orders. Hence, operationally, the migration activities are treated as managed (planned) service provisioning together with service termination for the old service.

12.2.3.2 Service Migration Design Example

The following service migration design activities need to take place in the design phase:

- define service migration process – joining the service provision and service termination process for the two services;
- customer network redesign – mainly configuration and IP address changes;
- define the tasks to be performed per customer site during migration.

In light of all the above, a service migration strategy and plan can be drawn up (please see following section for details). Service operational testing should be conducted before the start of the implementation phase.

12.2.3.3 Service Migration Strategy and Implementation Plan Example

Since the service is for corporate/business customers, project plans are created for each customer for scheduling the migration for each site together with engineer visits for installing and replacing the CPE. In this example, it might be wise to perform migration on one or two customers to trial the operational processes and work instructions for the migration. The new network configuration is prebuilt and tested with the new service profiles before the migration of each customer site. Parallel migration of customers can take place after successful customer trial migrations. Customer migration reviews take place biweekly to ensure the processes and technical prebuilding/preprovisioning work is working well and that issues are dealt with.

12.2.3.3.1 Entry and Exit Criteria

In this example, the most important gate review after business approval is gate 2. The entry criteria for gate 2 review include:

- customer redesigns are signed off;
- service migration processes and work instructions are signed off;
- service operational testing has been completed with no outstanding test exceptions;
- service migration strategy and implementation plan are signed off.

At the gate 2 review, the necessary resource availability for the service migration needs to be confirmed.

12.2.3.3.2 Implementation Plan

The implementation plan for the example is as shown in Figure 12.8.

After business approval and after the budget for service migration has been allocated, the formulation of the service migration strategy and plan, the redesign of customer networks and the definition of customer migration tasks and operational processes for customer migration can take place simultaneously. The customers should have been informed as part of the service withdrawal activities. Once the operational processes and the detailed work instructions have been produced and agreed, it is wise to perform operational service testing before customer service migration takes place. This will ensure that what has been defined is fit for purpose and will minimize the operational risks.

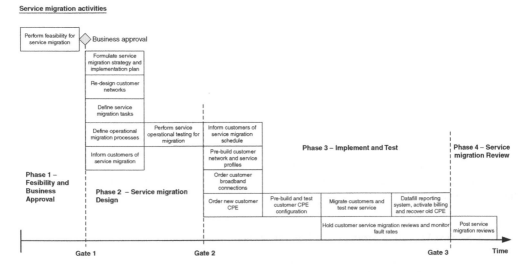

Figure 12.8   Example implementation plan for service migration.

As part of the implementation plan formulation, the customers to be migrated should be consulted and agree to their migration project plan. Other activities taking place after gate 2 approval include: prebuilding of customer network configuration and service profiles according to the new network designs that have been signed off; ordering the broadband access network connections and ordering CPE for the customers. Once all the CPE for a customer has arrived, the new configuration should be installed and tested with the new network configuration and service profiles that have been prebuilt. This should take place on a per customer basis and should be done before the engineers are sent out to customer sites.

Once the customer has been migrated successfully (i.e. all the installation tests have passed), the reporting systems are provisioned and the billing of the new service activated. After the migration of the first few customers, service migration reviews should take place to confirm that the operational processes and tasks defined are correct. Issues arising during the migration of a customer could be common to other customers. Therefore, reviewing these issues will benefit subsequent customer migrations.

A post-migration review is held 1 month after all service migration activities have been completed. This can coincide with any service withdrawal reviews if that proves to be efficient.

12.3 Service Termination

Service termination occurs when customers/end users decide not to use the service anymore. The scenario discussed here is the customers/end users terminating the service as well as the network. Under this circumstance, service termination can be viewed as reversing all the network and service provisioning activities. For business customers, these are normally planned activities where the network is disconnected for the service and the CPE recovered at all the customer sites. The service termination dates are agreed at the start of the termination process. For end users, normally, the service termination is as soon as possible. In both cases, the addresses for the final bills are to be confirmed with the customers/end users.

System functions for customers' and end-users' service termination is covered in Chapter 8 and the operational processes for service termination are covered in Chapter 9. However, I would like to highlight the following:

- CPE recovery from the customers/end users does not often happen after service termination. This leads to wasted assets and incurs unnecessary cost.
- The minimum term of contract is to be checked before termination and the correct charges applied at service termination.
- To improve the service and customer retention rate, one may consider asking customers/end users their reasons for leaving the service. This should help to keep the customers/end users and provide good feedback for the marketing department.
- Final bill payment collection may not be the easiest thing to do, especially if the customers/end users do not have a direct debit arrangement or other services with you.
- Network and system resources are often not released for reuse after customers/end users have terminated the service. This will lead to inefficient use of resources and network assets.

Appendix

New Network Technology Introduction Process

When introducing a new service, often new network technologies need to be used. The new network technology introduction process is designed to protect the existing network being operated. This process ensures that (i) the new network equipment introduced will not cause any damage to or have any adverse effect on the existing network and (ii) verifies that the new technology can be operated and managed effectively. This, in turn, minimizes the risk to the business when introducing new technologies.

Figure 1 illustrates the new network technology introduction process. This process can be used on any new technology introductions into the network and network upgrades, and is not confined to new technologies for new services. This process can be used in conjunction with the service design process detailed in Chapter 4. References are made to the service design process throughout this section where activities of the two processes coincide with each other. Different check point (i.e. decision point (DP)) reviews have been put in place to minimize the risks highlighted above. Depending on the organization of the project or program that has been put in place, representatives from both the design and network operational departments should be part of the review at various DPs.

One of the keys to the success of this process is to involve the operational areas at the earliest possible phase of the project. At the end of the day, they will be managing the technologies introduced and they need to ensure that the new equipment is operable, manageable and maintainable. There is no use having a piece of technology in the network that has fantastic features, but which is impossible to manage and maintain.

Phase 1: Project Initiation

In the project initiation phase, the object of the new technology has to be articulated and the scope of the project defined. As part of the scoping exercise, the requirements to be satisfied by the new technology are to be defined and agreed. Within the project initiation document, a list of entry and exit criteria for each phase should be listed.

Initial business approval is acquired either through the approval of initial business case from the service design process or an initial business case for the technology introduction itself. Without business buy-in, the project is not going to go anywhere.

Successful Service Design for Telecommunications Sauming Pang
© 2009 John Wiley & Sons, Ltd

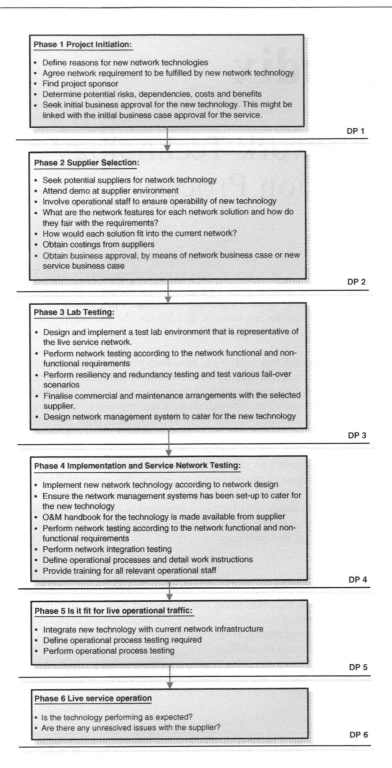

Phase 1 Project Initiation:

- Define reasons for new network technologies
- Agree network requirement to be fulfilled by new network technology
- Find project sponsor
- Determine potential risks, dependencies, costs and benefits
- Seek initial business approval for the new technology. This might be linked with the initial business case approval for the service.

DP 1

Phase 2 Supplier Selection:

- Seek potential suppliers for network technology
- Attend demo at supplier environment
- Involve operational staff to ensure operability of new technology
- What are the network features for each network solution and how do they fair with the requirements?
- How would each solution fit into the current network?
- Obtain costings from suppliers
- Obtain business approval, by means of network business case or new service business case

DP 2

Phase 3 Lab Testing:

- Design and implement a test lab environment that is representative of the live service network.
- Perform network testing according to the network functional and non-functional requirements
- Perform resiliency and redundancy testing and test various fail-over scenarios
- Finalise commercial and maintenance arrangements with the selected supplier.
- Design network management system to cater for the new technology

DP 3

Phase 4 Implementation and Service Network Testing:

- Implement new network technology according to network design
- Ensure the network management systems has been set-up to cater for the new technology
- O&M handbook for the technology is made available from supplier
- Perform network testing according to the network functional and non-functional requirements
- Perform network integration testing
- Define operational processes and detail work instructions
- Provide training for all relevant operational staff

DP 4

Phase 5 Is it fit for live operational traffic:

- Integrate new technology with current network infrastructure
- Define operational process testing required
- Perform operational process testing

DP 5

Phase 6 Live service operation

- Is the technology performing as expected?
- Are there any unresolved issues with the supplier?

DP 6

Figure 1   New network technology introduction process.

A cross-functional project team is formed to run the project. Team members should include:

- business owner of the project (normally the network director);
- network designers;
- network operational personnel;
- network second-line support personnel;
- network management system designer;
- a project manager;
- a member of the network and capacity planning team;
- service designer;
- operational process change manager.

Phase 1 Output and DP 1 Entry Criteria

These include:

- initial business case;
- PID;
- network functional and nonfunctional requirements document.

Major checks at DP 1 include:

- Is there business support for the introduction of the new technology?
- Is the project team in place?
- Has the scope of the project and requirements been signed off?

Phase 2: Supplier Selection

After initial business approval has been obtained, the supplier selection process can be initiated. In order that all the suppliers' solutions are measured equally, a scorecard matrix should be drawn up with priorities and weighting on all the requirements. It is common practice to send out a request for information (RFI) or request for proposal (RFP) to the suppliers for proposal responses in light of the requirements. Presentations and demonstrations from the suppliers will help the shortlisting process. It will be essential to involve network operational departments as part of the solution evaluation process. If there are test facilities available (either at the supplier's or in-house), it will be a good idea to get some hands-on experience with each of the solutions proposed. This will make the technology evaluation more accurate. This can take place as part of the feasibility phase (phase 2) of the service design process. Once the supplier has been selected, commercial negotiation starts and the final cost will be known. The final cost can be fed into the business case for the service in phase 3 of the service design process.

Phase 2 Output and DP 2 Entry Criteria

These include:

- supplier evaluation scorecards;
- cost information from suppliers;
- approve business case for new technology.

Major checks at DP 2 include:

- What are the network features for each network solution and how do they fare with the requirements?
- Is the cost prohibitive?
- Is the business case going to stand?

Phase 3: Laboratory Testing

Laboratory testing is where the network designers, system designers and network operational personnel can test the new technology to its limits (or to the limits imposed by the laboratory facilities). Most operators/service providers will have a reference model of the network where software upgrades or customer complaints are tested out. This laboratory must be a representative of the live service network, with all the necessary management systems in place. Otherwise, laboratory testing will become pointless and does not serve the purpose of limiting any potential damage the new technology may cause to the live network.

It is important to perform any tests that cannot be performed in the live network in the laboratory. For example, resiliency and fail-over testing may be OK in the laboratory environment, but in the live network it could be impossible to perform without causing a service outage. Also, there might be fault scenarios that cannot be simulated in the live network as they could be service affecting. All those types of test need to be conducted here.

The laboratory testing will also help the network designers to understand how best to utilize the features on offer and how best to fit the new technology into the existing network. The detailed network configuration can also be captured during laboratory testing. This should be part of the design phase (phase 3) of the service design process.

For systems designers, this is their chance to get any network information (e.g. SNMP traps and MIB file) for the network management systems and network provisioning system (if relevant). Working alongside the network operational personnel is an effective way to design the network management system in parallel. The necessary system configuration and set-up can also be captured.

From the laboratory testing, it should also be apparent how the new technology needs to be managed. This is a useful input into negotiating the maintenance agreement and any commercial arrangement that is being discussed with the supplier.

Phase 3 Output and DP 3 Entry Criteria

These include:

- signed-off network design;
- sign off network management system design;
- operational & maintenance (O&M) manual from suppliers available.

Major checks at DP 3 include:

- Does the new technology do what it says on the box?
- Is it operable?
- What pitfalls are there with this technology?
- If the technology does not perform as expected, should alternative solutions be considered?

Phase 4: Implementation and Service Network Testing

When the business is satisfied with the new technology performing functions as expected in the laboratory environment, then the project can proceed to implementation in accordance with the signed-off designs. It is best if the service network testing can be performed on the actual network that the service is going to be operated on. Otherwise, an isolated part of the network in the live environment will be a good alternative.

Some of the network functional and nonfunctional testing from the laboratory testing will be performed here again to ensure the network has been built correctly. As part of the implementation, network integration testing is to be performed. This confirms that the new technologies are stable and have no adverse effect on the existing network and that the management systems are in place for operability and maintainability. Service testing is also performed to ensure the service to be supported works as designed. The integration testing and service testing can be carried out as part of the testing in phase 4 of the service design process. It is also good practice to have network operational personnel conducting or witnessing these tests. This can form part of the training for operational personnel and help them to write their operational processes and detailed work instructions.

Phase 4 Output and DP 4 Entry Criteria

These include:

- network functional and nonfunctional test results;
- network integration test results;
- service network test results;
- issues log result from network and service network testing conducted;
- operational processes and detailed work instructions documents.

Major checks at DP 4 include:

- Are the network management systems set up and data-filled correctly?
- Are the SNMP traps from the network elements interpreted correctly and the alarms displayed with the right severity?
- Does the network equipment perform the functions as expected?
- Is the network equipment stable in the live network environment?
- Is the service to be supported working as designed?
- What are the exceptions from the testing phase? Do these issues need to be resolved before proceeding to the next stage?

Phase 5: Operational Readiness

To ensure network operations are ready to operate the new technology, operational readiness testing is conducted. The tests being conducted depend on the service supported. However, most network operational testing will concentrate on fault scenarios, alarm management, disaster recovery and the ability to fix the faults on the equipment and supplier support processes. Customer and service provisioning, billing, network upgrade and configuration management are also areas to be considered for testing. If this was a network upgrade with new technology without changing any service features supported, then service level testing needs to be conducted. The operational readiness testing can be part of the service operate and launch phase (phase 5) of the service design process.

Phase 5 Output and DP 5 Entry Criteria

These include:

- operational readiness test results;
- issues log results from testing conducted.

Major checks at DP 5 include:

- Are all operational processes and detailed work instructions complete?
- Has operational readiness testing been completed with all exceptions logged?
- Upon reviewing the exceptions, is the new technology fit for 'live' traffic?

Phase 6: Live Service Operation

The new technology process does not end until the new technology has been in service for 3 months. This is to ensure that all the issues/teething problems are resolved satisfactorily. A DP 6 review should be set up 3 months after the DP 5 review to confirm everything is in order.

Major checks at DP 6 include:

- Are there any major unresolved issues with the supplier?
- Is the new technology behaving as expected?
- Does part of the network or system design need to be reworked to accommodate unexpected behaviors from the new technology?

Glossary

3G	third generation
3GPP	3rd Generation Partnership Project

A

ABR	available bit rate
ADSL	asynchronous digital subscriber loop
AKA	authentication and key agreement
ATM	asynchronous transfer mode
AuC	authentication center
AV	attribute value

B

B2B	business to business
BHCA	busy hour call attempt
BS	base station
BSC	base station controller
BSS	base station subsystem (in the context of mobile services)
	business support system (in the context of system functions)
BTS	base transceiver station

C

CAPEX	capital expenditure
CBR	constant bit rate
CDMA	code division multiple access
CDRs	call detail records
CDV	cell delay variation
CHAP	challenge handshake authentication protocol
CLI	caller line identity (in the context of voice services)
	command line interface (in the context of network management)
CLR	cell loss ratio
CPE	customer premise equipment
CMIP	common management information protocol
CN	corresponding node
CoA	care-of address (for mobile IP)
CoS	class of service

Successful Service Design for Telecommunications   Sauming Pang
© 2009 John Wiley & Sons, Ltd

CRM	customer relationship management
CS	circuit switched
CTD	cell transfer delay

D

D-GK	directory gatekeeper
DiffServ	differentiated services (RFC 2475)
DNS	domain name service
DP	decision point
DSCP	DiffServ code point
DSL	digital subscriber loop
DSLAM	digital subscriber loop add–drop multiplex
DWDM	dense wavelength division multiplexing

E

EDGE	Enhance Data for Global/GSM Evolution
EF	expedited forwarding
EIR	equipment identity register
ENUM	telephone *num*ber *m*apping (to domain name, email address or URL)
eTOM	enhanced telecom operations map

F

FA	foreign agent
FTP	file transfer protocol

G

GATS	General Agreement on Trade in Services
GERAN	GSM/EDGE RAN
GGSN	gateway GPRS support node
GK	gatekeeper
GMSC	gateway mobile switching center
GoS	grade of service
GPRS	general packet radio service
GSM	global system for mobile communication
GUI	graphical user interface

H

HA	home agent
HLR	home location register
HSS	home subscriber server
HTTPS	hypertext transport protocol over SSL

I

ICANN	Internet Corporation for Assigned Names and Numbers
IETF	Internet Engineering Task Force
IGRP	interior gateway routing protocol
IKE	internet key exchange
IMS	IP multimedia subsystem
IMSI	international mobile subscriber identity
IntServ	integrated services (RFC 1633)
IP	internet protocol
IPSec	IP security
IPv4	IP version 4

IPv6	IP version 6
ISDN	integrated service digital network
ISP	internet service provider
IT	information technology

K

KPI	key performance indicator

L

LAN	local area network
LLU	local loop unbundling

M

MAC	move, add and change
MAPSec	mobile application part security
MCUs	multipoint control units
MDF	main distribution frame
MDT	mean downtime
ME	mobile equipment
MIB	management information base
MG	media gateway
MGC	media gateway controller
MN	mobile node
MPLS	multi-protocol label switching
MSC	mobile switching center
MSISDN	mobile station international subscriber directory number
MTBF	mean time between failures
MTTF	mean time to failure
MTTR	mean time to repair/ recover
MVNO	mobile virtual network operator

N

NDS	network domain security
NPD	new product development

O

OFCOM	Office of Communications
OLO	other license operator (also known as other network provider and alternative network provider in this book)
OPEX	operational expenditure
OSA	open service access
OSPF	open shortest path first
OSS	operational support system

P

PBX	private branch exchange
PDSN	packet data service node
PHB	per-hop-behavior
PID	project initiation document
PKI	public key infrastructure
PPP	point to point protocol
PS	packet switched
PSTN	public switch telephone network

Q

QoS	quality of service

R

RACI	responsible, approval, consulted and informed
RAN	radio access network
RIP	routing information protocol (RFC 1388)
RF	radio-frequency
RFI	request for information
RFP	request for proposal
RNC	radio network controller
RNS	radio network sub-system
ROI	return on investment
RSVP	*resource reservation protocol*
RTP	real-time transport protocol

S

SCP	signaling control points
SDH	synchronous digital hierarchy
SDP	session description protocol
SEG	security gateway
SGSN	service GPRS support node
SHTTP	secure hypertext transport protocol
SIGTRAN	signaling transport
SIM	subscriber identity module
SIP	session initiation protocol
SLA	service level agreement
SLG	service level guarantee
SMS	short messaging service
SMTP	simple mail transfer protocol
SNMP	simple network management protocol
SONET	synchronous optical network
SS7	Signaling System 7 (also known as C7)
SSL	secure socket layer
SSP	signaling service point
SWOT	strength weaknesses opportunities and threats

T

TLD	top-level domain

U

UAT	user acceptance testing
UE	user equipment
UBR	unspecified bit rate
UMTS	universal mobile telecommunication system
USIM	User SIM
UTRAN	UMTS RAN

V

VBR	variable bit rate
VBR-nrt	variable bit rate – non-real time
VBR-rt	variable bit rate – real time
VC	virtual circuit
VCI	virtual circuit identifier
VLR	visitor location register
VoIP	voice over IP

VP	virtual path
VPN	virtual private network
VV&T	validation, verification and testing

W

WAN	wide area network
WCDMA	wideband code division multiple access
WDM	wavelength division multiplexing
WiMAX	worldwide interoperability for microwave access

References

[1] *Oxford Advanced Learner's Dictionary*, 6th edn, Oxford: Oxford University Press; 2000.
[2] McCabe, J.D. *Network Analysis, Architecture and Design*, 3rd edn, Elsevier Inc., 2007.
[3] RFC 791 – Internet Protocol Definition (IPv4); Sept 1981.
[4] Newton, R. *The Project Manager*. Financial Times Prentice Hall; 2005.
[5] Kaseara, S. and Narang, N. *3G Mobile Networks Architecture, Protocols and Procedures*. McGraw-Hill; 2005.
[6] Dean, T. *Guide to Telecommunications Technologies*. Thomson Course Technology; 2003.
[7] Smith, C. and Collins, D. *3G Wireless Networks*. McGraw-Hill; 2006.
[8] Raab, S. and Chandra, M.W. *Mobile IP Technology and Applications*. Cisco Systems, Inc.; 2005.
[9] 3GPP specifications. www.3gpp.org/specs/specs.htm (last access September 2008).
[10] ISO/IEC 27001: 2005. Specification for Information Security Management; 2005.
[11] RFC 2865 Remote Authentication Dial In User Service (RADIUS); Jun 2000.
[12] Davies, G. *Designing and Developing Scalable IP Networks*. John Wiley & Sons, Ltd; 2004.
[13] Kasera, S. *ATM Networks Concepts and Protocols*. McGrawHill; 2007.
[14] RFC 2916 E.146 and DNS; Sept 2000.
[15] Van Helvoort, H. *SDH/SONET Explained in Functional Models*. John Wiley & Sons, Ltd; 2005.
[16] RFC 3031 Multiprotocol Label Switching Architecture (MPLS); Jan 2001.
[17] Karl, E.W. *Software Requirements*. Microsoft Press; 2003.
[18] IEEE Recommended Practice for Software Requirements Specifications; IEEE Std 830-1998, Oct 1998.
[19] Peters, J., Bhatia, M., Kalidindi, S., jee, S. *Voice over IP Fundamentals*, 2nd edn, Cisco Press; 2006.
[20] Lauesen, S. *Software Requirements Styles and Techniques*. Addison-Wesley; 2002.
[21] Zhu, H. *System Design Methodology: From Principles to Architectural Styles*. Butterworth-Heinemann; 2005.
[22] Lewis, W.E. *Software Testing and Continuous Quality Improvement*, 2nd edn, Auerbach Publications; 2005.
[23] ITU E.490. 1 Traffic Engineering – Measurement and Recording of Traffic: Overview Recommendations on Traffic Engineering; Jan 2003. www.itu.int/net/home/index.aspx (last access September 2008).
[24] Smith, C. and Gervelis, C. *Wireless Network Performance Handbook*. McGraw-Hill; 2003.
[25] Pezze, M. and Young, M. *Software Testing and Analysis: Process, Principles and Techniques*. John Wiley & Sons, Ltd; 2007.
[26] ITU E360.1 Framework for QoS routing and related traffic engineering methods for IP-, ATM-, and TDM- based multi-service networks; May 2002. www.itu.int/net/home/index.aspx (last access September 2008).
[27] Calder, A. and Watkins, S. *IT Governance: A Manager's Guide to Data Security and BS 7799/ISO 17799*, 3rd edn, Kogan Page; 2005.
[28] Bosworth, S. and Kabay, M.E. (eds) *Computer Security Handbook*. John Wiley & Sons, Ltd; 2002.
[29] ISO/IEC 27002:2005 Information technology – Security technique – Code of practice for information security management; 2005.

[30] Wasson, C.S. *System Analysis, Design, and Development: Concepts, Principles, and Practices.* John Wiley & Sons, Ltd; 2006.

[31] Enhanced Telecom Operations Map (eTom) Version 7.1; Jan 2007.

[32] Massam, P. *Managing Service Level Quality Across Wireless and Fixed Networks.* John Wiley & Sons, Ltd; 2003.

[33] Norris, M. *Mobile IP Technology for M-Business.* Artech House, Inc.; 2001.

[34] Chuan, M.C. and Zhang, Q. *Design and Performance of 3G Wireless Networks and Wireless LANs.* Springer Science + Business Media, Inc.; 2006.

[35] Balakrishnan, R. *Advanced QoS for Multi-Service IP/MPLS Networks.* John Wiley & Sons, Ltd; 2008.

[36] Minoli, D. *Voice over MPLS: Planning and Design Networks.* McGraw-Hill (Telecom Engineering); 2002.

[37] Gray, C.F. and Larson, E.W. *Project Management: The Managerial Process.* McGraw-Hill Higher Education; 2002.

[38] RFC 1633 Integrated Service in the Internet Architecture: An Overview (IntSev); Jun 1994.

[39] RFC 2212 Specification of Guaranteed Quality of Service; Sept 1997.

[40] RFC 2211 Specification of the Controlled-Load Network Element Service; Sept 1997.

[41] RFC 2475 An Architecture for Differentiated Services (DiffServ); Dec 1998.

[42] RFC 2598 An Expedited Forwarding PHB; Jun 1999.

[43] RFC 2597 Assured Forwarding PHB Group; Jun 1999.

[44] Gómez, G. and Sánchez, R. *End-to-End Quality of Service Over Cellular Networks.* John Wiley & Sons, Ltd; 2005.

[45] Camarillo, G., García-Martín, M.-A. *The 3G IP Multimedia Subsystem (IMS): Merging the Internet and the Cellular Worlds*, 2nd edn, John Wiley & Sons, Ltd; 2006.

[46] Kaaran, H., Ahtiainen, A., Laitinen, L., *et al. UMTS Networks: Architecture, Mobility and Services*, 2nd edn, John Wiley & Sons, Ltd; 2005.

[47] Ash, G.R. *Traffic Engineering and QoS Optimization of Integrated Voice and Data Networks.* Elsevier Inc.; 2007.

[48] Liotine, M. *Mission-Critical Network Planning.* Artech House Inc.; 2003.

[49] Mishra, A.R. *Fundamentals of Cellular Network Planning and Optimisation.* John Wiley & Sons, Ltd; 2004.

[50] Borroughs, L. and Judge, T. *IP in Mobile Networks.* BWCS; 2004.

[51] Judge, T. *OSS Guide for Telecom Service Providers and ISPS: Untangling the Threads.* BWCS; 2002.

[52] Ghys, F., Mampaey, M., Smouts, M. and Vaaraniemi, A. *3G Multimedia Network Services, Accounting, and User Profiles.* Artech House Inc.; 2003.

[53] Terplan, K. *OSS Essentials: Support System Solutions for Service Providers.* John Wiley & Sons, Inc.; 2001.

[54] TeleManagement Forum. Telecom Application Map (TAM), Version 2.4; Aug 2007.

[55] TeleManagement Forum. *SLA Management Handbook, Volume 2 – Concept and Principles*, Version 2.5. Reading: The Open Group; 2005.

[56] TeleManagement Forum. *SLA Management Handbook, Volume 3 – Services and Technology Examples*, Version 2.0. Reading: The Open Group; 2004.

[57] Pasricha, H. *Designing Networks with Cisco.* Charles River Media; 2004.

[58] RFC 2460 Internet Protocol version 6 (IPv6) Specification; Dec 1998.

[59] Van Bosse, J.G. and Devetak, F.U. *Signaling in Telecommunication Networks*, 2nd edn, John Wiley & Sons, Inc; 2006.

[60] RFC 4301 Security Architecture of the Internet Protocol; please also see RFCs 2402, 2403, 2405, 2410, 2411, 2412, 4303, 4835, 4306.

[61] Kreher, R. *UMTS Performance Management: A Practical Guide to KPIs for the UTRAN Environment.* John Wiley & Sons, Ltd; 2006.

[62] RFC 1661 The Point to Point Protocol; Jul 1994.

[63] RFP 1994 PPP Challenge Handshake Authentication Protocol (CHAP); Aug 1996.

[64] RFC 1035: Domain Names – Implementation and Specification; Nov 1987.

[65] RFC 3261 SIP: Session Initiation Protocol; Jun 2002.

[66] RFC 2327 SDP: Session Description Protocol; Apr 1998.

[67] RFC 3550 RTP: A Transport Protocol for Real Time Applications; Jul 2003.

[68] RFC 3605 Real Time Control Protocol attribute in Session Description Protocol; Oct 2003.

Index

Printed and bound by CPI Group (UK) Ltd, Croydon, CR0 4YY

27/10/2024

14580295-0002